Ocular Hypertension - New Advances

Edited by Felicia M. Ferreri

Published in London, United Kingdom

Ocular Hypertension - New Advances
http://dx.doi.org/10.5772/intechopen.111272
Edited by Felicia M. Ferreri

Contributors
Aida Abovyan, Alyona Lifanova, Ankit Shah, Artashes Zilfyan, Brandon W. Montoya, Daniel Laroche, Deepkumar Patel, Felicia M. Ferreri, Gian Franco Díez Cattini, Ildar Lutfarakhmanov, Imani Nwokeji, Karen Allison, Kevin Morabito Jr, Kevin Peng, Mark Disclafani, Owen Banda, Peter Mironov, Sohum Sheth, Thokozani Mzumara, Valentine Pavlov

First published in London, United Kingdom, 2024 by IntechOpen
IntechOpen is the global imprint of INTECHOPEN LIMITED, registered in England and Wales, registration number: 11086078, 167-169 Great Portland Street, London, W1W 5PF, United Kingdom

British Library Cataloguing-in-Publication Data
A catalogue record for this book is available from the British Library

Additional hard and PDF copies can be obtained from orders@intechopen.com

Ocular Hypertension - New Advances
Edited by Felicia M. Ferreri
p. cm.
Print ISBN 978-1-83769-458-7
Online ISBN 978-1-83769-457-0
eBook (PDF) ISBN 978-1-83769-459-4

For EU product safety concerns:
IN TECH d.o.o., Prolaz Marije Krucifikse Kozulić 3, 51000 Rijeka, Croatia,
info@intechopen.com or visit our website at intechopen.com.

Meet the editor

Felicia M. Ferreri graduated *summa cum laude* from the University of Messina, Italy, in 1998 and completed her ophthalmology residency at the Policlinico Universitario, Messina, in 2002. She interned at the Corneal Section of San Raffaele Hospital in Milan and the Pediatric Ophthalmology Diseases Section of Hospital Careggi in Florence. She spent research periods at Virginio del Rocio Hospital in Seville, San Carlos Hospital in Madrid, the Royal Bolton Hospital in Manchester, and Universidade Fluminense in Rio de Janeiro. She served as a co-investigator of many national and international clinical trials. Since 2002, she has been an Assistant Professor in Ophthalmology at the University of Messina. Her research interests are glaucoma, neuro-ophthalmology, pediatric ophthalmology, and cataract. She authored more than 50 scientific papers and edited two IntechOpen Books.

Contents

Preface

This Edited Volume *Ocular Hypertension – New Advances* is a collection of reviewed and relevant research chapters, concerning the developments within the Ocular Hypertension field of study.

The book includes scholarly contributions by various authors and was edited by a group of experts who are pertinent to ophthalmology. Each contribution comes as a separate chapter complete in itself but directly related to the book's topics and objectives.

The target audience comprises scholars and specialists in the field.

Intechopen Team

Chapter 1

Effect of Selective Laser Trabeculoplasty in Ocular Hypertension

Aida Abovyan and Artashes Zilfyan

Abstract

The aim of the study was to compare the results of using eye drops (a beta-blocker) with the results of selective laser trabeculoplasty in ocular hypertension. The retrospective study was conducted among treatment-naïve patients of two clinics in Yerevan, Armenia, during a period from 2019 to 2022. The eligible participants were divided into two groups as follows: group 1 (85 eyes) with patients who received SLT once and group 2 (52 eyes) where the patients have applied a single type of eye drops. The patients were allocated to the groups taking into account their preferences; this corresponded to random sampling. The IOP measures taken after 1, 3, 6, and 12 months showed reduction for more than 20% each time compared to baseline data. The mean IOP reduction was not significantly different in both treatment groups. SLT, as an effective and convenient technique allowing avoidance of the side effects of eye drops and an issue of low compliance, can be recommended as a method of the first choice in primary prevention of glaucoma.

Keywords: selective laser Trabeculoplasty, ocular hypertension treatment, selective laser Trabeculoplasty in ocular OHT, SLT effect, SLT in ocular hypertension

1. Introduction

Use of laser for intraocular pressure (IOP) reduction is dated to 1970s. In 1979, Wise and Witter suggested low-energy argon laser photocoagulation (ALT) and concluded that it was a successful procedure for the short-term IOP decrease [1].

The mechanism of ALT is ambiguous [2]. According to mechanical theory, laser thermal burns to trabecular meshwork (TM) lead to collagen contraction and local scarring, thus enhancing space for nearby tissues. In obedience to biological theory, thermal energy of ALT causes structural changes of TM by means of "altered cytokine secretion, matrix metalloproteinase induction, increased cell division, repopulation of burn sights, and macrophage recruitment". As a result, extracellular matrix is changed, and intraocular fluid outflow increases [2, 3].

In 1983, Anderson and Parrish found that pigmented cells of tissue could absorb applied radiation energy selectively, resulting in localized damage. This process was called selective photothermolysis (SP) [4].

SP had two main criteria. The first criterion was an intracellular chromophore with a higher optical absorption at the laser wavelength than in neighboring tissues, and the second criterion was that the period of laser radiation could not be longer than it is required for thermal diffusion into the tissue (thermal relaxation time) [5].

The ALT duration (~0.1 s) is longer than the thermal relaxation time of melanin (1 μs); it allows heat generated in pigmented cells to dissipate and damage collateral TM [5].

Selective Laser Trabeculoplasty (SLT) was introduced in 1995 by Latina and Park. They used a 532 nm Q-switched, frequency-doubled Nd:YAG laser that generated a shorter pulse duration (3 ns). It suits the abovementioned requirements of SP, preventing heat dissolution outside of pigmented TM cells and causing less surrounding damage [6].

In 2001, SLT received approval of Food and Drug Administration (FDA). According to Arora et al., 75,647 trabeculoplasties were performed in 2001 in the USA; this number was almost doubled, reaching 142,682 procedures in 2012 [7].

SLT lowers the IOP by increasing the outflow through the TM [8].

Histopathological study showed that SLT causes less damage to TM compared to ALT [9].

It is suggested that TM disruption could depend on energy dosage, as higher power SLT can lead to more severe TM damage than lower power SLT [10].

The mechanical and biological theories, which have been suggested to explain ALT's mechanism of action, do not entirely suit to SLT, as limited damage of tissues occurs to the TM in SLT, as thermal relaxation time of melanin is approximately 1 millisecond and the pulse duration of SLT laser is 3 nanosecond [5].

SLT lowers intraocular pressure (IOP) by improving aqueous outflow mainly by biological changes in the TM [5].

SLT has demonstrated the regulation of expression of genes responsible for cell motility, extracellular matrix production, membrane repair, and reactive oxygen species production [11].

In vitro studies have shown an increase in pro-inflammatory cytokine expression, including interleukin-1-alpha, interleukin-1-beta, tumor necrosis factor-alpha, and interleukin-8 post SLT [12].

These cytokines enhance stromelysin-1 expression (MMP-3), a matrix metalloproteinase, involved in TM extracellular matrix remodeling to increase aqueous outflow through the juxtacanalicular meshwork [13].

After SLT, increase in TM monocyte recruitment has been registered as a result of increased chemokine production [14].

Several studies demonstrated that local increase in endothelin-1 may lead to acute IOP rise after SLT (IOP spike); meanwhile, increase in lipid peroxide levels and reduction in antioxidant enzyme levels can occur due to the increased inflammatory response precipitated after laser [15, 16].

In vivo, monocytes improve aqueous outflow and, *in vitro*, facilitate Schlemm's canal permeability through the secretion of cytokines or directly phagocytosis of the debris within TM [15].

Kramer et al. conducted a study, which compared changes after SLT and ALT in the uveal meshwork part of the TM in autopsy eyes. After ALT, TM structures detected to be damaged with coagulative destruction, (ablation craters), while in SLT, minimal mechanical damage of tissues occurred. At the same time, cracking of intracytoplasmic pigment granules and the disruption of TM endothelial cells were observed [4]. Signs of destruction and scrolling of trabecular beams at the edges of laser burns appeared on tissues treated with energy of 2 mJ [7].

Alvarado JA et al. showed that SLT and prostaglandin (PGA) analogues may have a common pathway of action by causing intercellular junction disassembly in Schlemm's canal and TM cells in this way increasing aqueous permeability [17, 18].

SLT was evaluated as the first-line therapy by Melamed et al. and McIlraith et al. in prospective studies and demonstrated reduction of IOP by approximately 30% compared to baseline levels, which is comparable with prostaglandin efficacy [19–21].

According to Laser in Glaucoma and Ocular Hypertension (LiGHT) Trial, SLT was found to be a clinically safe and cost-effective procedure as an initial treatment of open-angle glaucoma (OAG) and ocular hypertension (OHT) at 3 years [22].

SLT is performed using topical anesthetic. The patient sits at the slit lamp; a single mirror gonioscopic lens (Latina lens) with coupling medium is used. Laser beam is focused on the pigmented part of TM [18].

A frequency doubled, Q-switched Nd:YAG laser is used during SLT. The wavelength is 532 nm. Pulse duration is 3 ns, and laser spot size is 400 μm. Pulse energy ranges from 0.4 to 1.4 mJ [5, 23].

The laser energy is initially set at 0.6 mJ. If cavitation bubbles ("champagne bubbles") appear, the energy is reduced by 0.1 mJ increments until the formation of bubbles is minimal, and treatment is continued at this energy level [5, 23].

Several studies compared treating different degrees of the TM with SLT to see how it influences IOP lowering [18].

Chen et al. had discovered that there is no difference between the application of SLT over 90° and 180° using 25 nonoverlapping laser spots per quadrant [24].

Other studies have shown that 180° and 360° laser is more effective than 90° laser. And the success rate is the highest over 360° SLT [25, 26].

In current practice, typical treatment parameters are 50–100 shots applied over 180°–360° with energy adjusted to 0.6–1.4 mJ and the end point of visible tissue reaction in the form of microbubbles [18, 23].

The indications for SLT include as follows:

- Newly diagnosed open-angle glaucoma (OAG) patients,
- OHT,
- OAG patients uncontrolled on medical treatment,
- OAG patients with likely or actual poor compliance or poor tolerance to medical treatment,
- patients with pseudoexfoliation or pigmentary glaucomas,
- patients who cannot afford glaucoma medications,
- patients who would like to reduce the number of glaucoma medications they are taking,
- patients who do not tolerate glaucoma medications [27, 28].

Current contraindications are as follows:

- Inflammatory/uveitic glaucoma;
- congenital glaucoma;
- poor visualization of the TM;
- active uveitis or history of uveitis (relative contraindication);
- traumatic glaucoma (relative contraindication);
- congenital or early childhood glaucoma (relative contraindication);
- primary or secondary angle-closure glaucoma (relative contraindication) [27, 28].

Contrary to the past theories, Ho et al. have reported that IOP is successfully decreased by SLT in eyes with primary angle closure and a patent iridotomy in case when there was a sufficient extent area of visible TM [29].

Some studies have shown that long-term use of topical medication can lead to conjunctival fibrosis and postoperative bleb dysfunction [30, 31].

Complications of SLT include elevated IOP, iritis, hyphema, macular edema, foveolar burn. and corneal haze [32].

2. Materials and methods

The aim of the study was to compare the results of using eye drops (a beta-blocker) with the results of selective laser trabeculoplasty in ocular hypertension.

The retrospective study was conducted among patients who visited Zilfyan Eye Care Clinic, Yerevan, Armenia, or Shengavit Medical Center, Yerevan, Armenia, during a period from 2019 to 2022. The clinical records of the patients were reviewed, and those that met the inclusion criteria were included into the study. Totally, 1473 records were reviewed.

The eligible records were divided into two groups as follows: group 1 (85 eyes) with treatment-naïve patients with ocular hypertension, who received SLT once and group 2 (52 eyes) where the patients with ocular hypertension have applied a single type of eye drops for the first time (Timolol 0.5%). The patients were allocated to the groups taking into account their preferences; this approached random sampling. The two groups were not significantly different by age and gender characteristics, which allowed minimizing the confounding factors.

Inclusion criteria were treatment-naïve patients with newly diagnosed ocular hypertension whose intraocular pressure exceeds 21 mmHg, no evidence of glaucomatous optic neuropathy and no glaucoma changes on Humphrey field analyzer (by the "SITA" standart algorithm), and who visited the clinic after treatment starting in 1, 3, 6 months, and a year.

Exclusion criteria were any type of glaucoma, any concomitant eye diseases, previous treatment for ocular hypertension (OHT), any previous eye surgery, and patients who missed a visit to the clinics after 1, 3, 6 months, or 1 year (incomplete results). In addition, those patients from SLT group, who had to use eye drops because of not reaching lower IOP level, were not included into the study.

In the first group, the SLT was done with Q-switched Nd:YAG laser wavelength 532 nm, with a spot size of 400 μm. The laser system was coupled to a slit lamp delivery system with a helium-neon laser (HeNe) aiming beam. The eye was anesthetized with topical eye drops (tetracaine hydrochloride 1%). The patient was seated in front of slit lamp. Latina goniolens with coupling medium (artificial tear gel) was used, and laser beam was focused on pigmented part of trabecular meshwork. Laser spot size was 400 μm. Hundred nonoverlapping shots (25 per quadrant) with pulse duration of 3 nanosecond were applied.

The laser energy varies within 0.8–1.3 mJ range. If cavitation bubbles formed, the energy was decreased by 0.1 mJ, and laser was done at that energy level. If no cavitation bubbles appeared, the pulse energy was increased by 0.1 mJ steps until "champagne bubbles" formation was observed, then decreased again by 0.1 mJ.

After the procedure, anti-inflammatory steroidal eye drops (dexamethasone 0.1%) were prescribed 4 times a day during 5 days. The patients were examined the next day afterSLT, in order not to miss the surge of IOP spike, which can occur after SLT.

The participants of the second group were using a B-blocker 2 times a day (every 12 hours).

In order to prevent poor compliance, the explanation of how and when to use eye drops was provided to the patients at each visit.

At baseline, visual acuity testing, IOP measurement (I-care manual rebound tonometer), slit-lamp examination, automated visual field (VF) testing (Humphrey Field Analyzer, the Swedish Interactive Threshold Algorithm 24-2 program), gonioscopy, CCT measurement, evaluation of optic disc, macula, and fundus were performed to all patients.

The IOP was measured with I-care manual rebound tonometer before and after eye drops or SLT at next day, 1 month later, 3 months later, 6 months later, and 1 year later.

In cases when target IOP was more than 10 mmHg on the next day after SLT (IOP spike), a 7-day course of a topical aqueous inhibitor was prescribed.

Two-sample T-test (Welch's T-test) was used to compare the clinical effect of one-time SLT and single type of eye drops. A p value of <0.05 was considered statistically significant.

3. Results

The results were calculated using online statistics calculators: www.calculator.net and https://www.statskingdom.com/

The baseline characteristics of the study participants include age, gender, and laterality.

Participant baseline characteristics are shown in **Table 1**. In the group of SLT, the average age of the patients was 64.43 years (± 10.46 years), with more female patients than male (21.15% male vs. 78.85% female).

Thirty-three patients out of 52 (63.46%) had bilateral OHT, and 19 patients (36.54%) had unilateral OHT.

In the eye drops group, the average age of the patients was 58.55 years (± 16.24 years), and again, the number of female patients was higher than male (32.35% male vs. 67.65% female). Eighteen patients out of total 34 (52.94%) had bilateral OHT, while 16 patients (47.06%) had unilateral OHT (**Table 1**).

The initial mean IOP was 25 mmHg ±2.89 (SD) in the group treated with SLT and 24.38 mmHg ±2.46 (SD) in the group treated with eye drops. High P-value (0.1833)

Characteristics	SLT group	Eye drops group
Age (years), Mean (SD)	64.43 ± 10.46	58.55 ± 16.24
Gender (patients), (%)	11 (21.15%)	11 (32.35%)
Male	41 (78.85%)	23 (67.65%)
Female		
Laterality (patients), (%)	33 (63.46%)	18 (52.94%)
Both eyes	7 (13.46%)	8 (23.53%)
Right eye	12 (23.08%)	8 (23.53%)
Left eye		

Table 1.
Baseline characteristics of SLT and eye drops groups.

indicates that the difference between the sample average of initial IOP in 2 groups was not big enough to be statistically significant (CI = 95%).

The median value was 24 in both groups, and mode was 23 and 22 in SLT group versus 22 in the group using eye drops (**Table 2**).

The mean IOP measured 1 month after intervention was 18.4 mmHg ± 3.85(SD) in the first group and 17.94 mmHg±3.97(SD) in the second group. The difference in treatment effect between the two groups 1 month after the treatment was not significant ($p = 0.5071$) (CI = 95%).

The median value was 19 and 18 and, and the mode was 16 and 15 in first and second groups, respectively (**Table 3**).

Three months after the intervention, the mean average slightly decreased to 18.19 mmHg±3.23 (SD) in the first group and 17.48 mmHg ± 3.82 (SD) in the second

Initial IOP	**SLT**	**Eye drops**
Number of eyes	85	52
Mean (average)	25	24.38
Median	24	24
Mode	23, 22 (each appeared 17 times)	22 (appeared14 times)
Largest	34	33
Smallest	22	22
Range	12	11
Standard Deviation	2.89	2.46
Variance	8.35	6.04
Sample Standard Deviation	2.90	2.48
Sample Variance	8.45	6.16

Table 2.
Results of IOP evaluation before SLT and eye drops usage.

IOP – after 1 month	**SLT**	**Eye drops**
Number of eyes	85	52
Mean (average)	18.49	17.94
Median	19	18
Mode	16 (appeared 12 times)	15 (appeared 7 times)
Largest	28	26
Smallest	10	10
Range	18	16
Standard Deviation	3.85	3.97
Variance	14.85	15.79
Sample Standard Deviation	3.88	4.01
Sample Variance	15.03	16.09

Table 3.
Results of IOP evaluation after 1 month of SLT and eye drops usage.

group. The treatment effect 3 months after intervention again was close in both groups (p = 0.2664, CI = 95%). The median was 18 and 17, respectively, and the mode was 18 in the first group and 17 and 14 in the second group (**Table 4**).

The IOP evaluation after 6 months showed approximately similar mean in the first and second groups, 18.49 mmHg±3.64 (SD) and 18.6 mmHg ±4.15 (SD), respectively (p = 0.1151, CI = 95%). The median IOP was 18 and the mode 17 in the first group and 19 and 23, respectively, in the second group (**Table 5**).

The mean IOP measured 1 year after the intervention was 19 mmHg ±3.4(SD) in the SLT group and 19.08 mmHg±4.17(SD) in the group using eye drops. No difference between the sample averages of 2 groups was found (p = 0.9068, CI = 95%). The medians were 19 and 20 and modes were 20 and 23, respectively (**Table 6**).

IOP – after 3 months	SLT	Eye drops
Number of eyes	85	52
Mean (average)	18.19	17.48
Median	18	17
Mode	18 (appeared 15 times)	17, 14 (each appeared 7 times)
Largest	27	27
Smallest	10	8
Range	17	19
Standard Deviation	3.23	3.82
Variance	10.44	14.6
Sample Standard Deviation	3.24	3.86
Sample Variance	10.55	14.88

Table 4.
Results of IOP evaluation after 3 months of SLT and eye drops usage.

IOP – after 6 months	SLT	Eye drops
Number of eyes	85	52
Mean (average)	18.49	18.6
Median	18	19
Mode	17 (appeared 12 times)	23 (appeared 7 times)
Largest	30	26
Smallest	11	10
Range	19	16
Standard Deviation	3.62	4.15
Variance	13.07	17.20
Sample Standard Deviation	3.64	4.19
Sample Variance	13.23	17.54

Table 5.
Results of IOP evaluation after 6 months of SLT and eye drops usage.

IOP – after 12 months	SLT	Eye drops
Number of eyes	85	52
Mean (average)	19	19.08
Median	19	20
Mode	20 (appeared 12 times)	23 (appeared 9 times)
Largest	29	28
Smallest	11	10
Range	18	18
Standard Deviation	3.406	4.13
Variance	11.6	17.04
Sample Standard Deviation	3.426	4.17
Sample Variance	11.74	17.39

Table 6.
Results of IOP evaluation after 12 months of SLT and eye drops usage.

Treatment effect is more visible when expressed in percentage comparing the initial results with follow up at certain points. Thus, 1 month after the intervention, the IOP evaluation showed 26.38 and 26.2% decrease in the first and second groups, respectively. Three months later, the IOP fell down by 27.21% in the SLT group and by 28.1% in the group using eye drops. Further, after 6 months, the IOP reduction was 26.01 and 24.06% in the first and second groups, respectively. Finally, the decrease in IOP 1 year later was 23.97% in the SLT group and 21.51% in the group using SLT.

4. Conclusion

The results of the study have shown that both SLT and eye drops are effective methods of ocular hypertension treatment. In both groups, the IOP measures taken after 1, 3, 6, and 12 months showed reduction for more than 20%.

Comparison of two methods demonstrated no statistically significant difference in treatment effects. Yet further research with larger sample sizes and longer follow-up time is advisable.

This study, in difference with other similar studies, focuses on treatment of OHP cases only, whereas other studies include also glaucoma cases. The value added of the study therefore is the exploration of effectiveness of glaucoma primary prevention.

SLT is proven to be an effective and convenient IOP reduction method [18]. It helps to solve a problem of poor compliance and side effects of eye drops [33]. Therefore, SLT can be offered as a method of the first choice in treatment of ocular hypertension.

Author details

Aida Abovyan* and Artashes Zilfyan
Shengavit Medical Center, National Institute of Eye Health after S. Avdalbekyan, Zilfyan Eye Care Centre, Yerevan, Armenia

*Address all correspondence to: aidabov@gmail.com

References

[1] Wise JB, Witter SL. Argon laser therapy for open-angle glaucoma. A pilot study. Archives of Ophthalmology. 1979;**97**(2):319-322. DOI: 10.1001/archopht.1979.01020010165017

[2] Filippopoulos T et al. Argon laser Trabeculoplasty. In: Medscape. 3 Mar 2023. Available from: https://emedicine.medscape.com/article/1844064-overview

[3] Gazzard G, Konstantakopoulou E, Garway-Heath D, Garg A, Vickerstaff V, Hunter R, et al. Selective laser trabeculoplasty versus eye drops for first-line treatment of ocular hypertension and glaucoma (LiGHT): A multicentre randomised controlled trial. Lancet. 2019;**393**(10180):1505-1516. DOI: 10.1016/S0140-6736(18)32213-X Epub 2019 Mar 9. [Erratum in: Lancet. 2019 Jul 6;394(10192):e1]

[4] Anderson RR, Parrish JA. Selective photothermolysis: Precise microsurgery by selective absorption of pulsed radiation. Science. 1983;**220**:524-527

[5] Alon S. Selective laser Trabeculoplasty: A clinical review. Journal of Current Glaucoma Practice. 2013;**7**(2):58-65. DOI: 10.5005/jp-journals-10008-1139 Epub 2013 May 9

[6] Latina MA, Park C. Selective targeting of trabecular meshwork cells: in vitro studies of pulsed and CW laser interactions. Experimental Eye Research. 1995;**60**:359-371

[7] Arora KS, Robin AL, Corcoran KJ, Corcoran SL, Ramulu PY. Use of various glaucoma surgeries and procedures in medicare beneficiaries from 1994 to 2012. Ophthalmology. 2015;**122**:1615-1624

[8] Shaw E, Gupta P. Laser Trabeculoplasty. [Updated 2023 Mar 10]. In: StatPearls [Internet]. Treasure Island (FL): StatPearls Publishing; 2023. Available from: https://www.ncbi.nlm.nih.gov/books/NBK578175/

[9] Kramer TR, Noecker RJ. Comparison of the morphologic changes after selective laser trabeculoplasty and argon laser trabeculoplasty in human eye bank eyes. Ophthalmology. 2001;**108**:773-779

[10] Sacks ZS, Dobkin-Bekman M, Geffen N, Goldenfeld M, Belkin M. Non-contact direct selective laser trabeculoplasty: Light propagation analysis. Biomedical Optics Express. 2020;**11**(6):2889-2904. DOI: 10.1364/BOE.390849

[11] Sarenac T, Bečić Turkanović A, Ferme P, Gračner T. A review of selective laser Trabeculoplasty: "the hype is real". Journal of Clinical Medicine. 2022;**11**(13):3879. DOI: 10.3390/jcm11133879

[12] Bradley JM, Anderssohn AM, Colvis CM, Parshley DE, Zhu XH, Ruddat MS, et al. Mediation of laser trabeculoplasty-induced matrix metalloproteinase expression by IL-1beta and TNFalpha. Investigative Ophthalmology & Visual Science. 2000;**41**:422-430

[13] Lee JY, Kagan DB, Roumeliotis G, Liu H, Hutnik CM. Secretion of matrix metalloproteinase-3 by co-cultured pigmented and non-pigmented human trabecular meshwork cells following selective laser trabeculoplasty. Clinical & Experimental Ophthalmology. 2016;**44**:33-42

[14] Alvarado JA, Katz LJ, Trivedi S, Shifera AS. Monocyte modulation of

aqueous outflow and recruitment to the trabecular meshwork following selective laser trabeculoplasty. Archives of Ophthalmology. 2010;**128**:731-737

[15] Guzey M, Vural H, Satici A. Endothelin-1 increase in aqueous humour caused by frequency-doubled Nd:YAG laser trabeculoplasty in rabbits. Eye (London, England). 2001;**15**:781-785

[16] Guzey M, Vural H, Satici A, Karadede S, Dogan Z. Increase of free oxygen radicals in aqueous humour induced by selective Nd:YAG laser trabeculoplasty in the rabbit. European Journal of Ophthalmology. 2001;**11**:47-52

[17] Alvarado JA, Iguchi R, Martinez J, Trivedi S, Shifera AS. Similar effects of selective laser trabeculoplasty and prostaglandin analogs on the permeability of cultured Schlemm canal cells. American Journal of Ophthalmology. 2010;**150**:254-264

[18] Garg A, Gazzard G. Selective laser trabeculoplasty: Past, present, and future. Eye (London, England). 2018;**32**(5):863-876. DOI: 10.1038/eye.2017.273 Epub 2018 Jan 5. [Erratum in: Eye (Lond).] 2020;34(8):1487

[19] Melamed S, Ben Simon GJ, Levkovitch-Verbin H. Selective laser trabeculoplasty as primary treatment for open-angle glaucoma: A prospective, nonrandomized pilot study. Archives of Ophthalmology. 2003;**121**(7):957-960

[20] McIlraith I, Strasfeld M, Colev G, Hutnik CM. Selective laser trabeculoplasty as initial and adjunctive treatment for open-angle glaucoma. Journal of Glaucoma. 2006;**15**(2):124-130

[21] Van der Valk R, Webers CA, Schouten JS, Zeegers MP, Hendrikse F, Prins MH. Intraocular pressure-lowering effects of all commonly used glaucoma drugs: A meta-analysis of randomized clinical trials. Ophthalmology. 2005;**112**(7):1177-1185

[22] Gazzard G, Konstantakopoulou E, Garway-Heath D, Adeleke M, Vickerstaff V, Ambler G, et al. Laser in glaucoma and ocular hypertension (LiGHT) trial: Six-year results of primary selective laser Trabeculoplasty versus eye drops for the treatment of glaucoma and ocular hypertension. Ophthalmology. 2023;**130**(2):139-151. DOI: 10.1016/j.ophtha.2022.09.009 Epub 2022 Sep 17

[23] Jha B, Bhartiya S, Sharma R, Arora T, Dada T. Selective laser Trabeculoplasty: An overview. Journal of Current Glaucoma Practice. 2012;**6**(2):79-90. DOI: 10.5005/jp-journals-10008-1111 Epub 2012 Aug 16

[24] Chen E, Golchin S, Blomdahl S. A comparison between 90 degrees and 180 degrees selective laser trabeculoplasty. Journal of Glaucoma. 2004;**13**(1):62-65. DOI: 10.1097/00061198-200402000-00012

[25] Nagar M, Ogunyomade A, O'Brart DP, Howes F, Marshall J. A randomised, prospective study comparing selective laser trabeculoplasty with latanoprost for the control of intraocular pressure in ocular hypertension and open angle glaucoma. The British Journal of Ophthalmology. 2005;**89**(11):1413-1417

[26] Shibata M, Sugiyama T, Ishida O, Ueki M, Kojima S, Okuda T, et al. Clinical results of selective laser trabeculoplasty in open-angle glaucoma in Japanese eyes: Comparison of 180 degree with 360 degree SLT. Journal of Glaucoma. 2012;**21**(1):17-21

[27] Nagar M. Selective laser trabeculoplasty. In: Garg A, Rosen E,

editors. Instant Clinical Diagnosis in Ophthalmology: Glaucoma. New Delhi: Jaypee Brothers Medical Publisher Ltd; 2009. pp. 298-307

[28] Damiji K. Selective laser trabeculoplasty: A better alternative. Survey of Ophthalmology. 2008;**53**(6):646-651

[29] Ho CL, Lai JS, Aquino MV, Rojanapongpun P, Wong HT, Aquino MC, et al. Selective lasertrabeculoplasty for primary angle closure with persistently elevated intraocular pressure after iridotomy. Journal of Glaucoma. 2009;**18**(7):563-566

[30] Broadway DC, Grierson O'BC, et al. Adverse effects of topical Antiglaucoma medication: II. The outcome of filtration surgery. Archives of Ophthalmology. 1994;**112**(11):1446-1454

[31] Goldstein MH, Silva FQ, Blender N, et al. Ocular Benzalkonium chloride exposure: Problems and solutions. Eye (London, England). 2022;**36**(2):361-368

[32] Song J. Complications of selective laser trabeculoplasty: A review. Clinical Ophthalmology. 2016;**10**:137-143. DOI: 10.2147/OPTH.S84996

[33] De Keyser M, De Belder M, De Groot V. Quality of life in glaucoma patients after selective laser trabeculoplasty. International Journal of Ophthalmology. 2017;**10**(5):742-748. DOI: 10.18240/ijo.2017.05.14

Chapter 2

The Importance of Gonioscopy and Angle Assessment in Ocular Hypertension

Daniel Laroche and Imani Nwokeji

Abstract

Ocular hypertension (OHT) is characterized by elevated intraocular pressure (IOP), without any visible optic nerve damage or visual field loss. The mean normal intraocular pressure is 15 mm Hg, and the mean intraocular pressure in patients with untreated glaucoma is 18 mm Hg. However, simply having ocular hypertension does not necessarily lead to the development of glaucoma, instead, it is deemed a considerable risk factor. An example is a person with thick corneas who may have no glaucomatous damage with an IOP of 24 mm Hg. Thus, early detection and management of OHT and corneal pachymetry are imperative to help detect higher risk patients with thinner corneas with ocular hypertension or glaucoma early. The Laroche Glaucoma calculator is an effective inexpensive method to detect patients with glaucoma, glaucoma suspects, and ocular hypertensive with a higher risk of thinner corneas and older age. With respect to the physiology of ocular hypertension, the anterior chamber angle anatomy serves a crucial role in the regulation of IOP. Gonioscopy is an important technique for examining the angle structures, which provides essential information regarding the status of the trabecular meshwork and how this can affect aqueous outflow. This chapter will further explore the anatomy and physiology of the anterior chamber angle, specific principles, techniques, and interpretation of gonioscopy, the significance of early detection as well as the management of OHT.

Keywords: ocular hypertension, intraocular pressure, glaucoma development, risk factors, gonioscopy

1. Introduction

Ocular hypertension is a condition that is specified by an increase in intraocular pressure without any visible optic nerve damage or visual field loss [1–5]. The prevalence of ocular hypertension is approximately 2.7–3.8% in the general population and the incidence increases with the onset of age [6]. Ocular hypertension may not always prevail in development but it is appraised as a significant risk factor regarding the progression of glaucoma [7].

2. Anatomy and physiology of anterior chamber angle and aqueous outflow

2.1 Components of the anterior chamber angle

The anterior chamber angle is the region where the cornea, iris, and sclera meet, and subsequently the aqueous humor drains from the eye. This angle is composed of multiple structures such as the trabecular meshwork, Sclemm's canal, iris, ciliary body, and scleral spur. Primarily the trabecular meshwork regulates the outflow of aqueous humor from the eye [8]. Specifically, the Schlemm's canal is characterized as the principal pathway for aqueous humor. The ciliary body produces aqueous humor and the iris controls the size of the pupil; as the iris regulates the amount of light entering the eye. While the scleral spur is a bony ridge that anchors the ciliary body and trabecular meshwork [8].

2.2 Role of the trabecular meshwork in aqueous humor outflow

The trabecular meshwork (TM) (**Figure 1**) is the most essential part of the anterior chamber angle as it regulates aqueous humor outflow. It is a specialized tissue located at the anterior portion of the angle that is responsible for the majority of the resistance to aqueous humor outflow. The TM contains a meshwork of collagen fibers and endothelial cells that function to filter the aqueous humor before it flows into Schlemm's canal [8, 9]. Essentially, Schlemm's canal is a circular canal located around the circumference of the cornea that behaves as a collector channel for the aqueous humor to drain into the episcleral venous system [9].

The aqueous humor is generated by the ciliary body and flows through the posterior chamber into the anterior chamber, where it circulates and nourishes the cornea as well as the lens [8]. The aqueous humor must perpetually flow out of the eye to maintain a stable IOP. The TM is imperative in regulating the outflow of aqueous humor. Aqueous humor outflow happens through the mechanism of two different pathways; the trabecular meshwork pathway and the uveoscleral pathway [10].

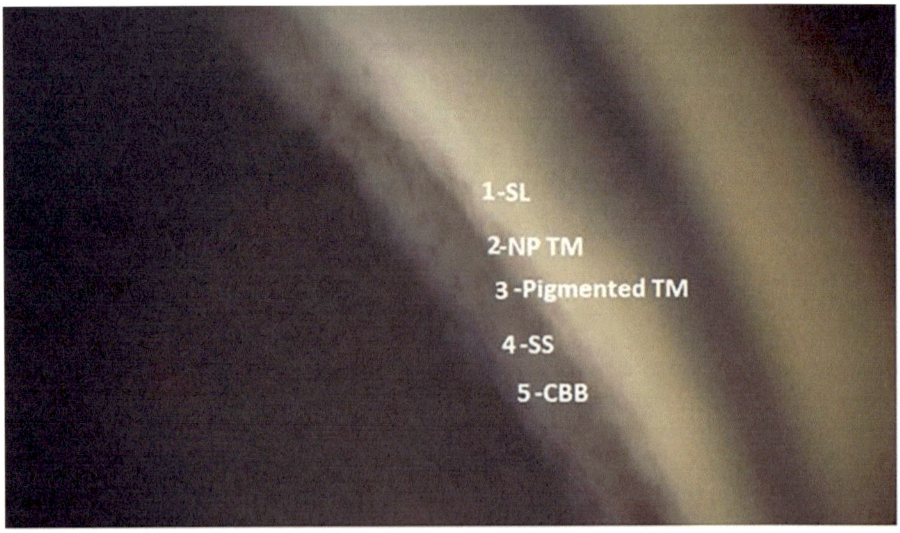

Figure 1.
Normal trabecular meshwork, 1-Schwalbe's Line, 1-Nonpigmented Trabecular Meshwork, 3-Pigmented Trabecular Meshwork, 4-Scleral spur, 5- Ciliary Body Band (Nidek GS1 Gonioscope).

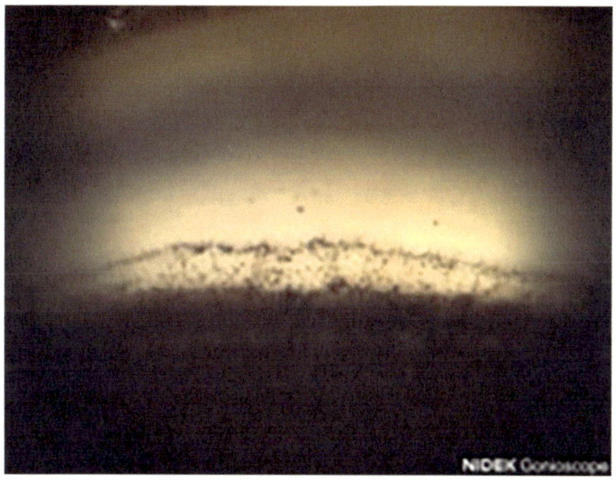

Figure 2.
Heavy pigmentation of inferior angle on (Nidek GS1 Gonioscope).

Essentially, the trabecular meshwork pathway is responsible for approximately 75% of aqueous humor outflow and is mostly regulated by the TM [11].

2.3 Factors that affect angle structures and obstruction of aqueous outflow

The regulation of intraocular eye pressure is an intricate process that involves the production and drainage of aqueous humor. The trabecular meshwork plays a vital role in the outflow of aqueous humor, and any obstruction or dysfunction of this structure can lead to an increase in IOP. A multitude of factors can affect the angle structures and outflow, including age, genetics, increased lens thickness, inflammation, trauma, medications, and systemic diseases [2]. The most common cause of outflow obstruction is the increased thickness of the lens in the eye. This compresses Schlemm's canal. The enlarged lens also leads to narrowing of the angle and increased iridozonular-iris rubbing, pigment liberation, and pigment obstruction of the trabecular meshwork (**Figure 2**) [12]. Cataract surgery expands the Schlemm's canal and is associated with the lowering of intraocular pressure [13]. Microinvasive glaucoma surgery bypasses the obstructed trabecular meshwork further lowering intraocular pressure [14].

3. Gonioscopy: technique and interpretation

3.1 Principles and technique of gonioscopy

Primarily, gonioscopy is a technique for examining the angle structure using a specialized lens. The procedure involves placing the gonioscope lens on the cornea to visualize the angle structures. The technique of gonioscopy involves an array of steps, first, there is patient preparation which is done before performing the gonioscopy, to adequately explain the procedure to the patient and to receive informed consent [15]. Furthermore, the eye is numbed with topical anesthetic drops to minimize any discomfort during the examination. Subsequently, the gonioscope lens is decided upon as there are multiple lenses available including Goldmann, Sussman, and Zeiss Lenses. This is selected based on the patient's anatomy, the examiner's preference as well as the level of detail required [16].

Then a coupling agent, typically a viscous gel or fluid is applied to the cornea to guarantee an appropriate amount of contact between the gonioscope lens and the corneal surface. As this enables optimal visualization of the anterior chamber angle structures of the eye. Furthermore, a gonioscope lens is placed accordingly, specifically on the cornea, securing that the entirety of the anterior chamber angle is visible through the lens. Throughout this process, the examiner may adjust as he or she sees fit [17].

This allows viewing of the angle by overcoming the total internal reflection at the cornea-air interface. A four-mirror lens is best to be able to perform indentation gonioscopy to see if there is an appositional vs. synechial closure of the angle. Indentation gonioscopy cannot be performed with a three-mirror gonioscope lens. The angle is graded according to the amount of visible trabecular meshwork and the extent of the angle opening. The grading system ranges from 0 to 4, with 0 indicating a closed angle and 4 which is characterized by a wide-open angle. The trabecular meshwork, Schlemm's canal, iris processes, and angle recesses are further evaluated. Assessment of these structures provides information regarding the angle width, presence of peripheral anterior synechiae, neovascularization as well as any pathological changes [18].

3.2 Interpretation of gonioscopic findings

Gonioscopy supplies beneficial information regarding the status of the trabecular meshwork and the degree of obstruction or dysfunction. The findings can help in the diagnosis and management of OHT and glaucoma. Anyone with an IOP of 18 mm Hg or higher should have a gonioscopy performed. For example, a narrow or closed angle can indicate a higher risk for angle-closure glaucoma, which requires prompt treatment to prevent vision loss. Heavier pigmentation in the inferior angle compared to the superior angle can be a sign of pigment liberation from lens iris rubbing contributing to elevated eye pressure [19].

3.3 Common pitfalls and challenges in gonioscopy

Ultimately, gonioscopy is a technique that requires expertise, however, some quite several pitfalls and challenges impact the accuracy as well as the reliability of the examination. This ranges from characteristics such as patient factors, technician factors, variability in angle structures, and the learning curve regarding the examiner. In regard to patient factors, many patients encounter challenges maintaining steady fixation throughout the procedure, thus leading to suboptimal visualization. Furthermore, small palpebral fissures, deep-set eyes, or substantial cornea edema can decrease the probability of attaining proper contact between the goniscope lens and the cornea [20].

Additionally, the physician's incorrectly applying the coupling agent or improperly positioning the gonioscope lens diminished the quality of the gonioscopic view. Hence the physician's proficiency in the technique plays a vital role in obtaining reliable and accurate results. Moreover, the variances in angle structures can primarily be attributed to age, race, and pathological conditions. Thus, acknowledging these variations and differentiating them from pathological changes proves to be difficult [18]. Lastly, examiners typically experience a learning curve concerning the gonioscopy findings, hence properly grading the angle necessitates experience as well as familiarity with different angle configurations [21].

4. Importance of gonioscopy and angle assessment in ocular hypertension

4.1 Role of gonioscopy in diagnosis and monitoring of ocular hypertension

Gonioscopy is pivotal in regarding the diagnosis and monitoring of ocular hypertension. Gonioscopic photos can be taken to educate patients about the findings of narrowing of the angle or increased pigmentation that may need to be addressed with an intervention. Medical therapy is often started to lower intraocular pressure. Currently, earlier cataract surgery and micro-invasive glaucoma surgery can lower intraocular pressure and reduce medication burden. Subsequently, cataract surgery has been noted to decrease the prevalence of angle closure [22].

4.2 Importance of angle assessment in predicting risk of glaucoma

Primarily, angle assessment through gonioscopy is necessary for predicting the risk of developing glaucoma in individuals with ocular hypertension. Previous research studies have indicated that narrow-angle or angle-closure configurations are associated with a higher risk of glaucoma development [23]. Patients with narrow angles require close monitoring and may be candidates for preventive treatment, such as laser peripheral iridotomy to prevent angle closure and consequent glaucomatous damage. Recently cataract surgery has been shown to be more effective than laser iridotomy in angle closure patients and this should be offered earlier to patients [24].

4.3 Relationship between angle status and response to treatment

The angle status of the patient has repercussions for the response to glaucoma treatment for patients with ocular hypertension. Studies in the past have demonstrated that angle status can influence the effectiveness of certain treatment modalities, such as medications or surgical interventions [25]. For example, for patients with open-angle configurations, medical therapy with topical glaucoma medications targeting the trabecular meshwork or uveoscleral outflow pathways prove to be more effective in diminishing intraocular pressure. These patients are also eligible for selective laser trabeculoplasty. Contrastingly, patients with synechial angle closure are not eligible for laser trabeculoplasty and may be less responsive to topical medications. They may need goniosynechialysis or trabeculectomy to migitigate angle obstruction and acquire sufficient intraocular pressure control [26].

5. Risk factors affecting angle structures and outflow

5.1 Age, race, and gender

Primarily, a multitude of risk factors can affect the angle structures and outflow in the eye. Age is an essential determinator as the anatomical and physiological changes that transpire with aging can implicate the angle. As individuals increase with age, the trabecular meshwork becomes reduced efficiency in draining the aqueous humor, provoking an increased risk for ocular hypertension and glaucoma. The size of the trabecular meshwork also decreases with age [27].

Race is another risk factor that has a profound impact, as several studies have demonstrated that specific racial groups, particularly those of African descent have a higher prevalence of narrow angles and angle-closure glaucoma. In comparison to patients of European descent which have an elevated risk of open-angle glaucoma. These differences demonstrated between patients of different races emphasize the influence that race has on predicting the risk of glaucoma. Black patients in the United States, Caribbean, and Africa have decreased access to healthcare and often present with more advanced glaucoma [28]. Additionally, individuals with more melinated skin also have brown irises with higher pigment density. In patients with glaucoma there is greater depigmentation of posterior pigment epithelium of the iris that can lead to obstruction of the trabecular meshwork [29]. This can be picked up on gonioscopy.

Furthermore, sex plays a crucial role in the angle structures and outflow. Previous studies indicated that hormonal differences between male and female individuals contribute to variations in the risk of developing glaucoma, however, the exact pathophysiology is not fully understood [30]. For example, many women are put at a higher risk of developing primary angle-closure glaucoma after menopause due to significant hormonal changes [30].

5.2 Systemic and ocular conditions

Specific systemic and ocular conditions can impact the angle structures and outflow, thereby enhancing the possibility of ocular hypertension and glaucoma. For example, systemic conditions such as diabetes and hypertension can affect the blood vessel in the eye, including those in the angle [31]. As well as conditions that cause inflammation, for instance, uveitis or pseudoexfoliation syndrome. These conditions can impair the angle and subsequently compromise the aqueous humor outflow [32]. Additionally, ocular conditions such as high myopia and patients that have had previous eye surgeries, particularly cataract surgery. Moreover, in regards to high myopia, the elongated shape of the eyeball can potentially cause mechanical changes in the angle, consequently resulting in increased resistance to aqueous outflow [33]. Other ophthalmic surgical procedures, mostly those involving the lens can alter the anatomical configuration of the anterior chamber and affect the angle [34].

5.3 Medications and lifestyle factors

Multiple medications can have a substantial effect on the angle structures and outflow dynamics. For example, medications that dilate the pupil can further narrow the angle and increase the risk of angle-closure glaucoma in susceptible individuals [35]. This is comparable to other systemic medications such as topiramate, which are used for migraines or epilepsy and have been associated with angle-closure glaucoma [36].

Furthermore lifestyle factors, particularly, smoking and excessive alcohol consumption can contribute to changes in the angle structures and outflow. Particularly, smoking has been associated with increased lens thickness, cataracts intraocular pressure, and a higher incidence of developing glaucoma [37]. Alcohol consumption, mostly heavy drinking can comprise the blood flow to the optic nerve and negatively impact the drainage of aqueous humor [38]. However, these factors are characterized as potential risk factors for angle structures and outflow but individual susceptibility can vary.

6. Clinical management of ocular hypertension

6.1 Importance of regular monitoring and risk stratification

Close observation and assessment of ocular hypertension are essential for promptly identifying the etiology and potential interventions to prevent vision impairments. The Laroche glaucoma calculator helps to identify patients who may be at increased risk based not only on IOP but also age and central corneal thickness [4]. Thus, the clinical management of ocular hypertension upholds the beliefs of preventative medicine and involves regular monitoring and risk stratification to ascertain patients with an elevated risk of developing glaucoma [39]. This typically includes periodic comprehensive eye examinations, measurement of intraocular pressure, assessment of the optic nerve, and evaluation of visual fields. Subsequently, risk stratification aids in recognizing individuals at a higher risk and prioritizing the management and follow-up of those patients [39].

6.2 Treatment options and goals

The main objective in managing ocular hypertension is to prevent or delay the onset of glaucoma and the associated vision loss. Treatment decisions are dependent on the level of intraocular pressure, risk factor, and the presence of structural or functional changes that are diagnostic of glaucoma. Early cataract surgery and lower intraocular pressure and reduce the need for medications [40]. In patients who have

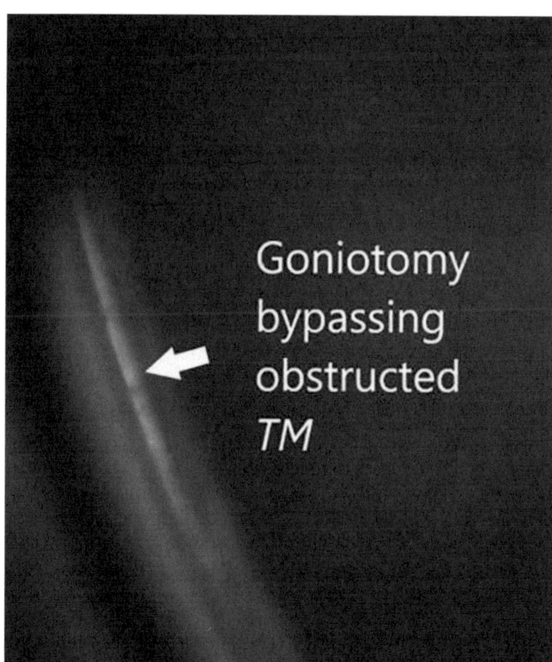

Figure 3.
Goniotomy performed at the time of cataract surgery with Sinskey hook bypassing obstructed trabecular meshwork at the time of cataract surgery.

developed mild to moderate glaucoma, early cataract surgery and microinvasive glaucoma surgery can lower intraocular pressure and reduce medication burden (**Figure 3**) [41]. Other treatment options include topical therapy and laser procedures, for example, selective laser trabeculoplasty of surgical interventions such as glaucoma filtering surgery.

However, the main objective of treatment is to reduce intraocular pressure to a level that diminishes the risk of glaucoma progression. The target intraocular pressure is specific to the patient as it is dependent on individual risk factors such as the extent of optic nerve damage, the patient's baseline intraocular pressure, and the other risk factors [39]. Clinicians approach lowering the intraocular pressure by decreasing aqueous humor production or increasing its outflow. Ultimately, the treatment plan is individualized to the patient considering medication compliance and potential adverse complications.

6.3 Role of gonioscopy in treatment decision making

Gionoscopy proves to be an essential contribution to the treatment plan in patients with ocular hypertension. As it aids in the assessment of the angle structures and provides beneficial insights regarding the status of the trabecular meshwork and the presence of angle abnormalities or pathology. In addition, it assists the clinician in determining a prognosis. For example, if gonioscopy discloses a narrow or closed angle it may suggest a higher risk of angle-closure glaucoma, which requires immediate intervention to prevent permanent vision loss. Heavy pigment in the angle due to pigment liberation from the thickened lens rubbing against the posterior iris may have one consider earlier cataract surgery and micro-invasive glaucoma surgery [41]. Furthermore, gonioscopy helps to identify specific angle abnormalities such as peripheral anterior synechiae or neovascularization which may have implications for treatment planning. The presence of neovascularization of the angle may require pan-retinal laser, anti-VEGF treatment and a glaucoma tube shunt [42].

Thereby, by providing detailed information about the angle structures, gonioscopy aids in individualizing treatment strategies and monitoring the response to treatment [43]. Clinicians can target specific angle abnormalities and optimize the management of ocular hypertension.

In summary, gonioscopy is pivotal in this process by providing valuable information about the angle structures and influencing treatment choices [43]. Thus by incorporating gonioscopy findings into the management plan, clinicians can optimize the care for patients and reduce the risk of glaucoma progression.

7. Future directions and challenges

7.1 Emerging technologies in angle assessment and imaging

The field of angle assessment is continuously evolving, undergoing advancements in imaging and the new emergence of technologies offers fruitful advancements in the diagnosis and management of ocular conditions [44]. The main objective of these new advancements in technology provide more detailed and comprehensive information concerning the anterior chamber angle, resulting in improved accuracy and precision in the evaluation of angle structures and outflow [45].

Moreover, the emerging technologies in angle assessment such as anterior segment optical coherence tomography provide the opportunity for high-resolution imaging of the anterior segment, which supplies detailed cross-sectional images of the angle structures [45]. This non-invasive imaging technique has the capacity to strengthen the evaluation of the trabecular meshwork, iris configuration, and angle dimensions, contributing beneficial observations for diagnosis and treatment planning [45].

An alternative sphere of advancement is the expansion of novel imaging modalities, including swept-source OCT and ultrasound biomicroscopy [46]. These techniques permit clinicians to visualize and analyze angle structures with improved depth penetration and resolution, facilitating complete analysis of the anterior chamber angle [45].

7.2 Addressing disparities in access to care and screening

Although advancements in technology demonstrate prospects for enhancing angle assessment, thus it is imperative to address disparities in access to care and screening [47]. Disparities in healthcare access, mostly amongst underserved populations, can postpone diagnosis and cause improper ocular conditions which include ocular hypertension and glaucoma [48, 49]. Thereby, community outreach programs, education campaigns, and collaborations between healthcare providers and community organizations aid in improving access to care and promoting early detection and intervention [50]. It is important for the industry to invest in media that often has a greater reach to higher-risk populations, more specifically Black patients as they disportiontely experience decreased access to care. It is also important to train more Black ophthalmologists and optometrists globally to meet the needs of Black communities around the world [51].

Remote screening programs, teleophthalmology, and digital health platforms can promote the early identification of patients at risk, granting time-efficient referrals for additional assessment and supervision [52]. Through the implementation of remote monitoring clinicians can monitor a patients intraocular pressure levels which allows for early detection of increases that can predispose patients to the onset of glaucoma [52].

7.3 Research priorities

Furthermore, determining research priorities is vital to propel advancements in the fields of gonioscopy and angle assessment. Numerous research priorities need to be addressed which include validation and standardization of grading systems, longitudinal studies, predictive models, cost-effectiveness analysis, patient-centered outcomes and quality-of-life measures [53]. Essentially, there is a demand for standardized and validated grading systems for evaluating the angle structures through gonioscopy. Consensus on terminology, definitions, and grading criteria will improve the consistency and conformity of findings across different studies [54]. Additionally, long-term studies are necessary to deepen comprehension between angle status, progression of ocular hypertension, and the development of glaucoma. Predictive models that incorporate multiple elements contributing to the risk factors can help identify individuals who are the most likely to benefit from early intervention [55].

Furthermore, assessing the cost-effectiveness of different screening and management strategies is essential for resource allocation and decision-making [56]. Examining the economic impact of implementing new technologies and interventions can help navigate healthcare policies and ensure the efficient allocation of healthcare resources [57]. Research should propel investigations regarding the development of patient-centered interventions and support systems. By confronting these research priorities, the field can make significant strides in optimizing the accuracy and effectiveness of angle assessment, mitigating differences in access to care, and improving patient outcomes in ocular hypertension and glaucoma management.

8. Conclusion

In conclusion, gonioscopy and angle assessment are imperative in managing ocular hypertension, a substantial risk factor for glaucoma and vision loss. Thus, comprehending the anterior chamber angle anatomy and physiology provides insights into outflow mechanisms and factors implicating angle structures. Gonioscopy provides visualizing and grading of the angle, assisting in diagnosis, monitoring, and treatment decisions. Risk factors including age, race, gender, and systemic conditions influence angle structures. Thus, consistent surveillance and treatment prioritize focusing on reducing intraocular pressure and preserving optic nerve health. Developing technologies and addressing disparities in care improve evaluation and access.

In summary, gonioscopy and angle assessment are imperative concerning ocular hypertension management. Gonioscopy assists with identifying the diagnosis and monitoring of ocular hypertension. This contributes sufficient information for the risk prediction of glaucoma and impacts treatment decisions [43]. With the incorporation of gonioscopy into the assessment of patients with ocular hypertension clinicians can obtain vital insights into the angling status, which permits appropriate risk stratification and personalized treatment approaches [15].

Conflict of interest statement

Dr. Laroche is a speaker for Sight Science, Johnson and Johnson, Nidek, and receives research support from Alcon. The information presented in this chapter is based on scientific evidence, medical knowledge, and professional experience without any financial or personal relationships that could potentially bias the content or influence the interpretation of the information provided. The main goal is to supply accurate, reliable, and unbiased information for the benefit of readers and the advancement of medical knowledge in the field of ocular hypertension and angle assessment.

Author details

Daniel Laroche[1,2]* and Imani Nwokeji[3]

1 New York Eye and Ear Infirmary of Mount Sinai, United States

2 Advanced Eyecare of New York, United States

3 CUNY School of Medicine, City College of New York, United States

*Address all correspondence to: dlarochemd@aol.com

References

[1] Tielsch JM, Katz J, Singh K, Quigley HA, Gottsch JD, Javitt J, et al. A population-based evaluation of glaucoma screening: The Baltimore eye survey. American Journal of Epidemiology. 1991;**134**(10):1102-1110. DOI: 10.1093/oxfordjournals.aje.a116013

[2] Kass MA, Hart WM Jr, Gordon M, Miller JP. Risk factors favoring the development of glaucomatous visual field loss in ocular hypertension. Survey of Ophthalmology. 1980;**25**(3):155-162. DOI: 10.1016/0039-6257(80)90092-2

[3] Abah ER, Mahmud-Ajeigbe AF, Sharief S, Chinda D, Jiya PY, Bob-Egbe O. Central corneal thickness and intraocular pressure of adult Nigerians: An assessment of Zaria community. West African Journal of Medicine. 2018;**35**(3):158-161

[4] Laroche D, Rickford K, Mike EV, Hunter L, Ede E, Ng C, et al. A novel, low-cost glaucoma calculator to identify glaucoma patients and stratify management. Journal of Ophthalmology. 2022;**2022**:5288726. DOI: 10.1155/2022/5288726

[5] Lei XW, Wei P, Li XL, Yang K, Lei JZ. Changes of structures of anterior chamber angle in rabbit chronic high intraocular pressure model. Zhonghua Yan Ke Za Zhi. 2009;**45**(10):892-897 Chinese

[6] Saboo US, Amparo F, Shikari H, Dana R. Prevalence of ocular hypertension and glaucoma in patients with chronic ocular graft-versus-host disease. Graefe's Archive for Clinical and Experimental Ophthalmology. 2016;**254**(5):923-928. DOI: 10.1007/s00417-016-3312-3. Epub 2016 Mar 12

[7] Nislawati R, Taufik Fadillah Zainal A, Ismail A, Waspodo N, Kasim F, Gunawan AMAK. Role of hypertension as a risk factor for open-angle glaucoma: A systematic review and meta-analysis. BMJ Open Ophthalmology. 2021;**6**(1):e000798. DOI: 10.1136/bmjophth-2021-000798

[8] Hann CR, Bentley MD, Vercnocke A, Ritman EL, Fautsch MP. Imaging the aqueous humor outflow pathway in human eyes by three-dimensional micro-computed tomography (3D micro-CT). Experimental Eye Research. 2011;**92**(2):104-111. DOI: 10.1016/j.exer.2010.12.010. Epub 2010 Dec 25

[9] Avtar R, Srivastava R. Modelling aqueous humor outflow through trabecular meshwork. Applied Mathematics and Computation. 2007;**189**(1):734-745. DOI: 10.1016/j.amc.2006.11.109

[10] Tamm ER, Braunger BM, Fuchshofer R. Intraocular pressure and the mechanisms involved in resistance of the aqueous humor flow in the trabecular meshwork outflow pathways. Progress in Molecular Biology and Translational Science. 2015;**134**:301-314. DOI: 10.1016/bs.pmbts.2015.06.007. Epub 2015 Jul 9

[11] Goel M, Picciani RG, Lee RK, Bhattacharya SK. Aqueous humor dynamics: A review. Open Ophthalmology Journal. 2010;**4**:52-59. DOI: 10.2174/1874364101004010052

[12] Laroche D, Capellan P. The Aging Lens and Glaucoma in persons over 50: Why early cataract surgery/refractive lensectomy and microinvasive trabecular bypass can prevent blindness and cure elevated eye pressure. Journal of the

National Medical Association. Aug 2021;**113**(4):471-473. DOI: 10.1016/j.jnma.2021.03.001. Epub 2021 Mar 27. PMID: 33781565

[13] Stamer WD, Braakman ST, Zhou EH, Ethier CR, Fredberg JJ, Overby DR, et al. Biomechanics of Schlemm's canal endothelium and intraocular pressure reduction. Progress in Retinal and Eye Research. 2015;**44**:86-98. DOI: 10.1016/j.preteyeres.2014.08.002. Epub 2014 Sep 16

[14] Saheb H, Ahmed II. Micro-invasive glaucoma surgery: Current perspectives and future directions. Current Opinion in Ophthalmology. 2012;**23**(2):96-104. DOI: 10.1097/ICU.0b013e32834ff1e7

[15] Narayanaswamy A, Vijaya L, Shantha B, Baskaran M, Sathidevi AV, Baluswamy S. Anterior chamber angle assessment using gonioscopy and ultrasound biomicroscopy. Japanese Journal of Ophthalmology. 2004;**48**(1):44-49. DOI: 10.1007/s10384-003-0004-4

[16] Barbero S. The Alvarez and Lohmann refractive lenses revisited. Optics Express. 2009;**17**(11):9376-9390. DOI: 10.1364/oe.17.009376

[17] Esporcatte BL, Vessani RM, Melo LA Jr, Yanagimori NS, Bufarah GH, Allemann N, et al. Diagnostic performance of optical coherence tomography and nonspecialist gonioscopy to detect angle closure. Journal of Current Glaucoma Practice. 2022;**16**(1):53-58. DOI: 10.5005/jp-journals-10078-1354

[18] Sadagopan KA, Periasamy P. A new lens-holder to facilitate slit-lamp bio-microscopic-aided fundus and gonioscopic examination. Oman Journal of Ophthalmology. 2021;**14**(2):115-116. DOI: 10.4103/ojo.ojo_383_20

[19] Laroche D, Rickford K, Sinon J, Brown A, Ng C, Sakkari S. Preventing blindness from glaucoma with patient education, the NIDEK GS-1 gonioscope, lensectomy and microinvasive glaucoma surgery. Journal of the National Medical Association. 2023;**115**(2):175-185. DOI: 10.1016/j.jnma.2023.01.014. Epub 2023 Feb 24

[20] Varadaraj V, Sengupta S, Palaniswamy K, Srinivasan K, Kader MA, Raman G, et al. Evaluation of angle closure as a risk factor for reduced corneal endothelial cell density. Journal of Glaucoma. 2017;**26**(6):566-570. DOI: 10.1097/IJG.0000000000000669

[21] Phu J, Wang H, Khuu SK, Zangerl B, Hennessy MP, Masselos K, et al. Anterior chamber angle evaluation using gonioscopy: Consistency and agreement between optometrists and ophthalmologists. Optometry and Vision Science. 2019;**96**(10):751-760. DOI: 10.1097/OPX.0000000000001432

[22] Hu CC, Lin HC, Chen CS, Kuo NW. Reduction in admissions of patients with acute primary angle closure occurring in conjunction with a rise in cataract surgery in Taiwan. Acta Ophthalmologica. 2008;**86**(4):440-445. DOI: 10.1111/j.1600-0420.2007.01066.x. Epub 2007 Nov 17

[23] Weinreb RN, Aung T, Medeiros FA. The pathophysiology and treatment of glaucoma: A review. Journal of the American Medical Association. 2014;**311**(18):1901-1911. DOI: 10.1001/jama.2014.3192

[24] Azuara-Blanco A, Burr J, Ramsay C, Cooper D, Foster PJ, Friedman DS, et al. Effectiveness of early lens extraction for the treatment of primary angle-closure glaucoma (EAGLE): A randomised controlled trial.

Lancet. 2016;**388**(10052):1389-1397. DOI: 10.1016/S0140-6736(16)30956-4

[25] Burr J, Azuara-Blanco A, Avenell A, Tuulonen A. Medical versus surgical interventions for open angle glaucoma. Cochrane Database of Systematic Reviews. 2012;(9):CD004399. DOI: 10.1002/14651858.CD004399.pub3

[26] Harasymowycz PJ, Papamatheakis DG, Ahmed I, Assalian A, Lesk M, Al-Zafiri Y, et al. Phacoemulsification and goniosynechialysis in the management of unresponsive primary angle closure. Journal of Glaucoma. 2005;**14**(3):186-189. DOI: 10.1097/01.ijg.0000159131.38828.85

[27] Chen Z, Sun J, Li M, Liu S, Chen L, Jing S, et al. Effect of age on the morphologies of the human Schlemm's canal and trabecular meshwork measured with swept-source optical coherence tomography. Eye (London, England). 2018;**32**(10):1621-1628. DOI: 10.1038/s41433-018-0148-6. Epub 2018 Jun 19

[28] Olawoye O, Kizor-Akaraiwe N, Pons J, Sarimiye T, Washaya J, Hughes S, et al. Clinical characteristics and stage at presentation of glaucoma patients in Sub-Saharan Africa. Journal of Glaucoma. 2022;**31**(9):717-723. DOI: 10.1097/IJG.0000000000002068. Epub 2022 Jun 27

[29] Bashkatov A, Genina E, Koblova E, Tuchin V. Estimation of melanin content in iris of human eye. Saratov Journal of Medical Scientific Research. 2008;**5688**:302-311. DOI: 10.1117/12.593651

[30] Tehrani S. Gender difference in the pathophysiology and treatment of glaucoma. Current Eye Research. 2015;**40**(2):191-200. DOI: 10.3109/02713683.2014.968935. Epub 2014 Oct 6

[31] Shoshani Y, Harris A, Shoja MM, Arieli Y, Ehrlich R, Primus S, et al. Impaired ocular blood flow regulation in patients with open-angle glaucoma and diabetes. Clinical & Experimental Ophthalmology. 2012;**40**(7):697-705. DOI: 10.1111/j.1442-9071.2012.02778

[32] Yanagi M, Kawasaki R, Wang JJ, Wong TY, Crowston J, Kiuchi Y. Vascular risk factors in glaucoma: A review. Clinical & Experimental Ophthalmology. 2011;**39**(3):252-258. DOI: 10.1111/j.1442-9071.2010.02455

[33] Bontzos G, Plainis S, Papadaki E, Giannakopoulou T, Detorakis E. Mechanical optic neuropathy in high myopia. Clinical and Experimental Optometry. 2018;**101**(4):613-615. DOI: 10.1111/cxo.12548

[34] Yamaguchi T, Negishi K, Yuki K, Saiki M, Nishimura R, Kawaguchi N, et al. Alterations in the anterior chamber angle after implantation of iris-fixated phakic intraocular lenses. Journal of Cataract & Refractive Surgery. 2008;**34**(8):1300-1305. DOI: 10.1016/j.jcrs.2008.04.033

[35] Lachkar Y, Bouassida W. Drug-induced acute angle closure glaucoma. Current Opinion in Ophthalmology. 2007;**18**(2):129-133. DOI: 10.1097/ICU.0b013e32808738d5

[36] Quagliato LB, Barella K, Abreu Neto JM, Quagliato EMAB. Topiramate-associated acute, bilateral, angle-closure glaucoma: Case report. Arquivos Brasileiros de Oftalmologia. 2013;**76**(1):48-49. DOI: 10.1590/S0004-27492013000100014

[37] Wang D, Huang Y, Huang C, et al. Association analysis of cigarette

smoking with onset of primary open-angle glaucoma and glaucoma-related biometric parameters. BMC Ophthalmology. 2012;**12**:59. DOI: 10.1186/1471-2415-12-59

[38] Braunger BM, Fuchshofer R, Tamm ER. The aqueous humor outflow pathways in glaucoma: A unifying concept of disease mechanisms and causative treatment. European Journal of Pharmaceutics and Biopharmaceutics. 2015;**95**:173-181. DOI: 10.1016/j.ejpb.2015.04.029

[39] Maier PC, Funk J, Schwarzer G, Antes G, Falck-Ytter YT. Treatment of ocular hypertension and open angle glaucoma: Meta-analysis of randomised controlled trials. BMJ. 2005;**331**(7509):134. DOI: 10.1136/bmj.38506.594977.E0. Epub 2005 Jul 1

[40] Mansberger SL, Gardiner SK, Gordon M, Kass M, Ramulu P, Ocular Hypertension Treatment Study Group. Cataract surgery lowers intraocular pressure and medication use in the medication group of the ocular hypertension treatment study. American Journal of Ophthalmology. 2022;**236**:53-62. DOI: 10.1016/j.ajo.2021.07.008. Epub 2021 Jul 17

[41] Laroche D, Oseni J, Nkrumah G, Ng C. Clear lensectomy with Hydrus stent in black and afro-Latino patients: A 1-year retrospective study. Journal of Ophthalmology. 2022;**2022**:8011745. DOI: 10.1155/2022/8011745

[42] Senthil S, Dada T, Das T, Kaushik S, Puthuran GV, Philip R, et al. Neovascular glaucoma—A review. Indian Journal of Ophthalmology. 2021;**69**(3):525-534. DOI: 10.4103/ijo.IJO_1591_20

[43] Baskaran M, Perera SA, Nongpiur ME, Tun TA, Park J, Kumar RS, et al. Angle assessment by EyeCam, goniophotography, and gonioscopy. Journal of Glaucoma. 2012;**21**(7):493-497. DOI: 10.1097/IJG.0b013e3182183362

[44] Weinreb RN, Moghimi S. Advances in ocular imaging. Asia-Pacific Journal of Ophthalmology. 2019;**8**(2):97-98. DOI: 10.22608/APO.201984

[45] Shinoj VK, Hong XJ, Murukeshan VM, Baskaran M, Tin A. Progress in anterior chamber angle imaging for glaucoma risk prediction - A review on clinical equipment, practice and research. Medical Engineering & Physics. 2016;**38**(12):1383-1391. DOI: 10.1016/j.medengphy.2016.09.014. Epub 2016 Oct 24

[46] Mao Y, Flueraru C, Chang S, Popescu DP, Sowa MG. Imaging of biomedical turbid tissue using catheter swept-source optical coherence tomography. In: 2010 IEEE International Workshop on Medical Measurements and Applications, Ottawa, ON, Canada. 2010. pp. 151-157. DOI: 10.1109/MEMEA.2010.5480212. Available from: https://ieeexplore.ieee.org/document/5480212

[47] Piyasena MMPN, Murthy GVS, Yip JLY, Gilbert C, Zuurmond M, Peto T, et al. Systematic review on barriers and enablers for access to diabetic retinopathy screening services in different income settings. PLoS One. 2019;**14**(4):e0198979. DOI: 10.1371/journal.pone.0198979

[48] Wu AM, Shen LQ. Racial disparities affecting black patients in glaucoma diagnosis and management. Seminars in Ophthalmology. 2023;**38**(1):65-75. DOI: 10.1080/08820538.2022.2152702. Epub 2022 Dec 1

[49] Altangerel U, Nallamshetty HS, Uhler T, Fontanarosa J, Steinmann WC, Almodin JM, et al. Knowledge about glaucoma and barriers to follow-up

care in a community glaucoma screening program. Canadian Journal of Ophthalmology. 2009;**44**(1):66-69. DOI: 10.3129/i08-175

[50] Meade CD, Menard J, Martinez D, Calvo A. Impacting health disparities through community outreach: Utilizing the CLEAN look (culture, literacy, education, assessment, and networking). Cancer Control. 2007;**14**(1):70-77. DOI: 10.1177/107327480701400110

[51] Nelson R. Telemedicine and telehealth: The potential to improve rural access to care. AJN, American Journal of Nursing. 2017;**117**(6):17-18. DOI: 10.1097/01.NAJ.0000520244.60138.1c

[52] Nangalia V, Prytherch DR, Smith GB. Health technology assessment review: Remote monitoring of vital signs--current status and future challenges. Critical Care. 2010;**14**(5):233. DOI: 10.1186/cc9208. Epub 2010 Sep 24

[53] Dy SM, Herr K, Bernacki RE, Kamal AH, Walling AM, Ersek M, et al. Methodological research priorities in palliative care and hospice quality measurement. Journal of Pain and Symptom Management. 2016;**51**(2):155-162. DOI: 10.1016/j.jpainsymman.2015.10.019. Epub 2015 Nov 17

[54] Desmond T, Tran V, Maharaj M, Carnt N, White A. Diagnostic accuracy of AS-OCT vs gonioscopy for detecting angle closure: A systematic review and meta-analysis. Graefe's Archive for Clinical and Experimental Ophthalmology. 2022;**260**(1):1-23. DOI: 10.1007/s00417-021-05271-4. Epub 2021 Jul 5. Erratum in: Graefe's Archive for Clinical and Experimental Ophthalmology. 2021 Sep 28

[55] Thompson SG, Higgins JP. Treating individuals 4: Can meta-analysis help target interventions at individuals most likely to benefit? Lancet. 2005;**365**(9456):341-346. DOI: 10.1016/S0140-6736(05)17790-3

[56] Guindo LA, Wagner M, Baltussen R, Rindress D, van Til J, Kind P, et al. From efficacy to equity: Literature review of decision criteria for resource allocation and healthcare decisionmaking. Cost Effectiveness and Resource Allocation. 2012;**10**(1):9. DOI: 10.1186/1478-7547-10-9

[57] Hoomans T, Severens JL. Economic evaluation of implementation strategies in health care. Implementation Science. 2014;**9**:168. DOI: 10.1186/s13012-014-0168-y

Chapter 3

Risk Factors for Increased Intraocular Pressure and Ophthalmic Complications during Robot-Assisted Laparoscopic Prostatectomy

Ildar Lutfarakhmanov, Alyona Lifanova, Peter Mironov and Valentine Pavlov

Abstract

Robot-assisted laparoscopic prostatectomy (RALP) is the most effective treatment option for prostate cancer. Special conditions of the operation affect intraocular pressure (IOP). The purpose of this review was to systematize new data on changes in IOP during RALP, to review the ophthalmic complications related to the robot-assisted approach, and to suggest measures to avoid such issues. A systematic search for articles of the contemporary literature was performed in PubMed database for complications in RALP procedures focused on positioning, access, and operative technique considerations. Several complications in RALP procedures can be avoided if the surgical team follows some key steps. Adequate patient positioning must avoid skin, peripheral nerve, and muscle injuries, and ocular and cognitive complications mainly related to steep Trendelenburg positioning in pelvic procedures. The robotic surgical team must be careful and work together to avoid possible complications. This review offers the first assessment of perioperative changes in IOP and ophthalmic complications during RALP and several steps in surgical planning to reach this goal. Further studies with a longer follow-up period are necessary to determine the clinical efficacy and safety of various types of general anesthesia.

Keywords: prostate cancer, radical prostatectomy, robotic surgery, Trendelenburg position, intraocular pressure

1. Introduction

Minimally invasive surgery, including robot-assisted laparoscopic approaches, has been promoted in a hope that their use would reduce surgical complications. Since the Intuitive da Vinci system (Intuitive Surgical Inc., Sunnyvale, CA, USA) was approved for use in 2000, robotic surgery has increasingly become the standard procedure for the management of a broad range of common surgical procedures from general and colorectal surgery to urogynecological procedures [1]. The use of robotic surgery accounted for only 1.8% of all general surgery procedures in 2012, and this increased to 15.1% by 2018 [2]. Robotic surgery is attractive for several reasons, such as laparoscopic surgery, it offers smaller incisions, a lower risk of infection, a shorter hospital stays, and a shorter convalescence than its open counterpart. In addition, unlike laparoscopic surgery, robotic approach has the advantage of increasing surgical dexterity due to the increased degrees of freedom of the instruments. Robot-assisted laparoscopic radical prostatectomy (RALP) is now in widespread use for the management of localized prostate cancer. With the increasing popularity, frequency, and acceptance of the RALP procedure, an awareness of unique intra- and postoperative complications is heightened. There are few specific reports of the limitations and complications of RALP, including that of increases in intraocular pressure (IOP). RALP requires specific body condition, in which the patients must be placed in a steep (23–45°) Trendelenburg position (sTp). Gravity allows the abdominal viscera to be pulled away from the pelvic cavity, resulting in a clearer operating field. However, this positioning may lead to complications, with several ocular complications having been reported since the inception of robotic-assisted surgery. The risk of perioperative complications is increased by incorrect patient positioning, inadequate fixation, or even a long time in the proper patient positioning.

This review outlines general procedures most likely to develop damaging IOP levels and their causative factors, the effect of anesthetic agents and techniques on IOP, recent scientific evidence highlighting the significance of perfusion changes during surgery, key aspects of postoperative visual loss and management approaches for high-risk patients presenting for surgery, and to assess the ophthalmic complications related specifically to the RALP and to suggest measures of how to avoid them.

2. Postoperative vision loss

2.1 Incidence of postoperative vision loss

Getting blind after recovery from general anesthesia in patients who undergo RALP is a very rare (0.02–0.1%) but catastrophic complication of robotic surgery [3, 4]. The first description of perioperative visual loss (POVL) in Medline was published in 1950 [5]. A serious ophthalmic consequence, namely the retinal detachment attributed to sTp, was first reported in 1952 [6]. The penetration of robotic technology in various surgical fields increased ophthalmic complications. In recent years, cases of POVL with non-ophthalmic surgery have been reported, but the exact incidence of POVL is unknown because the data come largely from retrospective studies and case reports [7, 8]. The incidence of POVL following non-ocular surgery has been estimated to be as low as 0.0002% and as high as 0.2% [9]. The number of cases reported POVL registry yearly has been decreasing from a peak in 2000 and has been dropping since. It is probably due to a reduction in operative times along with intraoperative blood transfusion [10]. Although the incidence is extremely low, the prognosis is extremely poor [11].

2.2 Etiology of postoperative vision loss

The most common neuro-ophthalmologic causes of POVL are ischemic optic neuropathies (ION), followed by central retinal arteria occlusion [12, 13]. The etiology of POVL is multifactorial and intricately interrelated with patient, anesthetic, and surgical factors [14]. The integrity of the delicate structures of the eye that mediate vision is dependent on the IOP. Ocular perfusion pressure (OPP) depends on the difference between mean arterial pressure (MAP) and IOP, so intraoperative IOP elevation is a risk factor for postoperative blindness. Yet, IOP acts to compress the vessels within the globe akin to a Starling resistor and is a key component that determines the OPP. Venous congestion and decreased optic nerve perfusion in the absence of cerebrovascular and ocular circulatory autoregulation in an anesthetized patient contribute to the development of ION, leading to POVL [15]. Multiple factors have been proposed as risk for intraoperative IOP elevation and ocular perfusion imbalance, including the head positioning [16]. sTp combined with pneumoperitoneum caused an increase in IOP, reduced ocular perfusion, and possibly POVL. These rapid fluctuations in IOP and perfusion play a role in the pathogenesis of the visual field defects and associated ocular morbidity that frequently complicate otherwise uneventful surgeries [17]. When IOP is significantly elevated in the setting of glaucoma or prolonged sTp, it is important to maintain stable ocular perfusion by increasing MAP or decreasing IOP. The potentially negative vascular occlusive effects of elevated IOP are more likely to affect patients who experience episodes of decreased OPP due to systemic hypotension. This may result from the hypotensive effects of anesthesia or episodes of intraoperative hypotension. Given the unique patient's sTp during RALP, ocular complications may be more likely to occur secondary to physiologic changes that occur within the eye itself.

The retina is one of the most metabolically active tissues in the body, and its functional integrity is dependent on an adequate blood supply with retinal function linearly related to the OPP. Retinal cell death has been demonstrated at OPP below 50 mmHg. Marked elevations of IOP (up to 4–5 times the normal value) with consequent borderline retinal and optic disk perfusion pressures occur for prolonged periods during many procedures, including RALP, especially, with their demand for sTp and/or hypotensive anesthesia, can induce IOP changes and ocular perfusion imbalance. The exact etiology of such outcomes is multifactorial, but ocular hypoperfusion plays a significant and frequently avoidable role. Those with preexisting compromised ocular blood flow are especially vulnerable to intraoperative ischemia, including those with hypertension, diabetes, atherosclerosis, or glaucoma. However, overly aggressive management of MAP and IOP may not be possible given a patient's comorbidity status, and it potentially exposes the patient to risk of catastrophic choroidal hemorrhage. Anesthetic management significantly influences the pressure changes in the eye throughout the perioperative period. Strategies to safeguard retinal perfusion, reduce the ischemic risk, and minimize the potential for expulsive bleeding must be central to the anesthetic techniques selected [17].

2.3 Risk factors for postoperative vision loss

Numerous risk factors for POVL have been identified that include older patients with elevated baseline IOP, patients with hypertension, diabetes, obesity, anemia, vascular disease, increased blood viscosity, and patients who smoke, as well as patients who experience intraoperative hypotension, blood transfusion, lower colloid use during fluid administration or prolonged surgical times, and patients who are positioned on horseshoe-shaped headrests [18]. Patients with angle-closure glaucoma are also at high risk for ocular injury even during short procedures [19].

In 2007, a case of a 62-year-old patient who developed ION with complete bilateral loss of vision after a robotic-assisted procedure lasting 6 hour 35 minutes was first reported [20]. ION was described in a 58-year-old man after laparoscopic sigmoidectomy lasting more than 6 hours. It was postulated that the patient suffered from hypotension in conjunction with an acute increase in IOP due to prolonged sTp [21]. Another mean operation time from skin incision to fascial closure reported was 105 min (range, 55–300) [22]. In this study, the mean was 5.46 hour$\pm$60.3 minutes, but similar to the 5.3 $\pm$ 1.0 hour reported earlier [23].

Although definitive evidence is lacking, the quantity of blood loss may be a risk for the occurrence of ION. The mean estimated blood loss (EBL) reported from 1500 consecutive cases was 111 mL (range, 50–500 mL) [22]. In a 62-year-old patient who developed ION, the EBL was 1200 mL [20]. In one study, the mean EBL was 350 $\pm$ 343 mL, with no cases of ION. On the other hand, one patient became blind after laparoscopic prostatectomy even without observed hypotension, hemodilution, metabolic disorders, or extreme blood loss, and baseline IOP and time spent in sTp were identified as the only factors predicting an increase in IOP [13]. Recently, a rare case of non-arteritic anterior ION in 58-year-old female patient following robotic-assisted hysterectomy was presented. On fundoscopy optic disc edema and splinter hemorrhages at the optic disc edges were observed. Fluorescein angiography showed hypofluorescence of the optic disc in the early phases due to filling delay followed by hyperfluorescence with leakage from disc capillaries in the late phases of the angiogram. So, this case of ION is an uncommon cause of POVL after robotic surgery [24]. The case of a 34-year-old female who underwent an uneventful laparoscopic hysterectomy and suffered from complete POVL following the operation was recently presented. Operating time was 174 minutes, and EBL was 75 mL. No cerebral hemorrhage or ischemia was detected on imaging. Funduscopic exam revealed no structural abnormalities. The following morning, she reported mild light perception. Later that night, she reported a partial return of visual acuity and was discharged home. At her two-week postoperative visit, her vision had returned to baseline [25].

3. Robotic-assisted laparoscopic prostatectomy

3.1 IOP changes during minimally invasive surgery

IOP is regulated by aqueous humor production, aqueous humor drainage, autoregulation and control of choroidal blood volume, vitreous humor volume, and extraocular muscle tone [17]. While the production of aqueous humor is stable, the outflow of aqueous humor to the venous system may be affected by choroidal blood volume, vitreous humor volume, and extraocular muscle tone. Many surgical procedures require pneumoperitoneum, where the air is insufflated into the abdominal cavity and subsequent head-down position, where the patient is positioned supine on the table with the head tilted below the feet at an angle of roughly 16-degree, and up to 25–40-degree in sTp, both can cause hemodynamic alterations that may influence IOP increasing venous congestion. After induction of anesthesia, there is an initial IOP decrease from baseline (IOP is reduced significantly more by propofol compared to volatile anesthetics) followed by a slight increase after insufflation of pneumoperitoneum. Afterward, the venous congestion increases both central venous pressure (CVP) and IOP. The latter increases in a time-dependent manner: mean IOP doubling within 60 minutes and, in 25% of cases, tripling within 120 minutes. This IOP increase may be exacerbated by the pneumoperitoneum-induced increase in partial pressure of carbon dioxide ($PaCO_2$) [23]. Indeed, the carbon dioxide (CO_2) insufflation leads to an increase in IOP due to decreased venous return and increased episcleral venous pressure. An increase in CM3, which is then, in turn, transmitted to the episcleral veins and from these to the capillaries and arterioles, also contributes to decreased optic nerve and ocular perfusion. The choroidal expansion may also lead to elevation [26]. This increased IOP then leads to decreased OPP. IOP, however, normally plateaus after about 30–60 minutes and decreases after return to the supine position. As the pneumoperitoneum can cause an increase in intraoperative IOP, one study compared IOP following extraperitoneal CO_2 insufflation and intraperitoneal CO_2 insufflation for the laparoscopic procedure. The change in intraoperative IOP was not statistically significant in either group, or there was no significant difference in intraoperative IOP change between the groups [27].

sTp is frequently used during minimally invasive surgery. IOP changes during laparoscopic or robotic hysterectomy conducted in the sTp were prospectively evaluated in 10 female patients with no history of ocular pathology [28]. There was a statistically significant trend of increasing the IOP from baseline to the second hour of sTp. The IOP remained significantly elevated once the patient was returned to the supine position.

Changes, in IOP depending on different positioning, were compared in patients without eye disease undergoing laparoscopic operations requiring sTp or reverse Trendelenburg tilt. A significant decrease in IOP was observed after the establishment of anesthesia irrespective of the combination of anesthetic agents employed. Elevation in IOP after abdominal insufflation with CO_2 to create pneumoperitoneum was observed in all patients with a trend toward a greater increase in IOP with standard sTp compared to a reverse Trendelenburg tilt, which was associated with a slight reduction in IOP in several patients. These changes were reversed in patients when intra-abdominal pressure (IAP) was not more than 14 mmHg, and operative time was not beyond 90 minutes. None of these required treatment on follow-up with resolution of the increase in all instances. These findings suggest that sTp and prolonged length of surgery are more important factors in attaining more dangerous elevations of IOP than the standard pneumoperitoneum induction for routine cases [29].

Current surgical robotics platforms, starting with the da Vinci® Si™ and certainly with the da Vinci® Xi™, allow one to easily obviate the risks associated with lithotomy positioning by performing RALP in supine. A majority of surgeons continue to perform robotic prostatectomy in lithotomy, although nearly two-thirds have access to the Xi system. It remains the standard approach to RALP despite advances in robotics that have allowed for greater flexibility in positioning and operative planning. The first survey of surgeons' perspectives on RALP positioning has shown that surgeons have not been adopting supine positioning. Only 22% of all respondents used supine positioning, and only half had considered it. The biggest reason for position choice appears to be the convenience to the surgical team, and this may be related in part to training inertia. Given the significant physiologic changes associated with positioning, the modern robotic surgeon should be familiar with positioning options offered by contemporary technologies, and surgeon and surgical team education efforts should be aimed at overcoming hesitation due to practice-inertia and toward greater creativity [30].

3.2 Risk factors for increasing IOP during sTp

Multiple intraoperative factors affect the IOP increase during pneumoperitoneum in the sTp: end-tidal carbon dioxide ($EtCO_2$), CVP, MAP, peak airway pressure (PAP), transperitoneal absorption of CO_2, intra-abdominal pressure (IAP), and duration of surgery (DoS) (**Table 1**). A combination of these factors, possibly together with abnormal self-regulation in the posterior portion of the optic nerve, prothrombotic trends, and other specific patient factors, can lead to enough decreased oxygen supply to the optical nerve to cause ischemic injury. Additionally, there may be a potential link between prolonged increase in ICP from sTp and pneumoperitoneum and short-term postoperative cognitive impairment [38].

With the aim to explore different trends of IOP change patterns and its related factors, IOP measurements were performed at different time points in four clinical common (lithotomy, lateral, prone, and supine) surgical positions during the pelvic or abdominal surgeries using laparoscopic techniques [39]. The surgical lithotomy position after the CO_2 pneumoperitoneum was changed to have different angles with head-low and foot-high positions. After 10 minutes of position change, IOP tended to rise. The reasons were that: (1) the thoracic pressure increased due to diaphragmatic elevation after the CO_2 pneumoperitoneum, (2) the anesthesiologist increased the ventilation volume/min to maintain end-tidal carbon dioxide ($EtCO_2$) in a reasonable range and the airway pressure elevation was transmitted to the thoracic cavity, increasing CVP, (3) the low head position also increased the volume of blood returning to the heart, which increased the venous pressure in the head and face, causing obstruction of venous return in the eye and head, resulting in the pressure increase in episcleral vein, and leading to pressure increase in IOP. Postural position for the surgery played a role in IOP change. After postural changes, the body maintains hemodynamic stability through a series of complex regulatory mechanisms, including the self-regulation system, venous and arterial systems, and neural reflexes. However, this regulatory mechanism was weakened under anesthesia so healthcare workers needed to understand the pattern of IOP changes in different surgical positions.

The negative effect of sTp was shown when compared IOP peaks in patients with healthy eyes who underwent RALP in sTp versus patients who underwent "open" or laparoscopic access in a supine position. A statistically significant increase in IOP was observed during RALP using sTp [40].

A systematic review of the studies on both elective patients and healthy non-anesthetized volunteers in the spinal, neurosurgical, and urological fields was identified, which explored the changes in IOP according to patient positioning, all reported significant rises in IOP in both head-down and prone positioning g, and the strongest effects were seen in those patients placed in combined head-down and prone position. Rises in IOP were time-dependent in all studies. Patients undergoing laparoscopic colorectal surgery in a prolonged head-down position were likely to experience raised IOP, and thus were at risk of POVL [41].

In another study, using univariate mixed effects models, PAP, MAP, $EtCO_2$, and DoS were significant predictors of the IOP increase, whereas age, BMI, EBL, volume of fluid administered, mean airway pressure, and desflurane concentration were not predictive. Surgical duration and $EtCO_2$ were the only significant variables predicting

First author, year	Awad 2009 [31]	Molloy 2011 [13]	Hoshikawa 2014 [32]	Molloy 2014 [33]	Yoo 2015 [34]	Blecha 2017 [26]	Demasi 2017 [35]	Goel 2020 [36]	Shirono 2020 [37]
Country	USA	USA	Japan	USA	Korea	Germany	Italy	India	Japan
Procedure	RALP	Laparoscopic	RALP	RALP and gynecologic laparoscopic	RALP	RALP	RALP	RALP	RALP
sTp angle, °	25	32–37	23	32–40	29	45	30	>45	NA
Carboxyperitoneum pressure, mmHg	15	NA	NA	14–15	8–20	15	15	15	NA
Total operative time	142 (105–210) min	3 h	4.57 ± 0.03 h	2.20 ± 0.56 h	113 ± 26	218 (120–357) min	143.56 ± 7.98 min	150 ± 41.7 min	265 (129–487) min
Estimated blood loss, mL	80 (45–155)	250 (50–600)	364 ± 196	159.8 ± 146.2	440 ± 288	NA	213.2 ± 98.6	NA	275 (0–1650)
Intravenous fluid, mL	2000 (1600–3100)	2500	NA	2058.7 ± 620.8	1264 ± 358	880 (550–2200)	1545 ± 174	NA	NA
Factor(s)	MAP PAP $EtCO_2$ DoS	DoS	DoS	BMI Age	IAP	MAP PAP	Angle of tilt	Obstruction in aqueous outflow	DoS

Note: NA, not available.

Table 1.
Factors determining increased IOP in sTp.

changes in IOP during stable and prolonged sTp. Indeed, the continued absorption of intraperitoneal CO_2, resulting in increased $PaCO_2$ and leading to vasodilation in the choroid plexus and an increase in IOP. Positive association between PAP and IOP throughout surgery, but not an increase over time was also shown. The proposed mechanism is that an increase in IOP leads to an increase in CVP, which can reduce the outflow of intraocular fluid through the episcleral veins and increase IOP [31].

Although prior studies have reported that pneumoperitoneum may increase IOP, it is not clear whether this increase is related to the effects of pneumoperitoneum or to the sTp. One study aimed to evaluate the potential fluctuations of IOP during colorectal laparoscopic surgery in two groups of patients: those with and those without sTp. In all the patients, standard pneumoperitoneum (<14 mmHg) induction led to a mild rise in IOP. The patients with Trendelenburg positioning showed a greater increase than the patients without it, but IOP evaluation 48 hours after surgery showed no substantial differences between the two groups. At the multivariate analysis, no potential predictors of increased IOP during surgery were identified. Thus, the patient's position during surgery may represent a stronger risk factor for IOP increase than pneumoperitoneum-related IAP [42].

The effect of sTp on IOP was examined during RALP. The highest IOP values were reached at sTp and intraperitoneal insufflation measurement time. Resistance index of the central retinal artery values was different, while impedance index of the central retinal vein values remained similar when patients were supine and awake and were anesthetized. Despite a long time of stay in sTp, the risk of ophthalmic complications was low [43].

sTp and pneumoperitoneum increased IOP directly proportional to the angle of tilt of the Trendelenburg position. Furthermore, the proportional lowering of OPP was directly correlated to the angle of tilt of the Trendelenburg position from the induction of anesthesia until the end of the procedure, but there is no agreement on the exact effects of MAP and IOP on blood flow in optic nerve [35].

Body mass index (BMI) was significantly correlated with IOP levels in patients, undergoing RALP performed with the sTp at baseline, 30, 60, and 90 minutes, and at the final time points when the surgery was finished. When the IOP levels were compared between the patients with BMI of 35 kg/m^2 and lower and the patients with BMI higher than 35 kg/m^2, the higher BMI patients had significantly higher levels of IOP. Patients' age was also positively correlated with IOP [33]. In contrast, one study demonstrated that age, BMI, DoS, and sTp did not affect IOP. Univariate mixed effects models showed PAP and MAP to be significant predictors for IOP increase [26]. Data from a study support that IOP significantly increases in a time-dependent manner after sTp in anesthetized patients undergoing RALP. Multiple perioperative factors are believed to be involved in controlling the increase in IOP during surgery, but there was no significant relationship between the changes in IOP and the age and BMI in this cohort [44]. In participants without a history of eye disease or eye surgery who underwent laparoscopic prostatectomy, bowel, and hysterectomy surgical procedures in sTp for a minimum of 120 minutes, the IOP increased from a range of 9 to 28 mmHg in the supine position to 25 to 54 mmHg at 120 minutes in sTp. About half cases were followed up through 3.5 hours of surgery in sTp. Ending IOP in the supine position was statistically significantly higher than baseline IOP. In 26% of cases, IOP tripled within 2 hours in sTp. Ocular perfusion pressure dropped below IOP in this greater than 30 BMI kg/m^2 patient population. There was a significant correlation of increase in mean IOP as time progressed. Several patients complained of blurred vision for a period following surgery, but POVL was not present [13].

Despite no patient experienced any ocular complications related to IOP increase, including ION, IOP was noted to increase in a time-dependent fashion. Along with previous reports, this suggests that longer operation times may induce substantially more risk for harmful IOP increases [32]. In the present study, IOP increased when patients were in sTp, and it was thereafter elevated in a time-dependent manner during sTp. In addition, the console time significantly affected the increase in IOP during RALP at a cut-off of 4 hours. So, to prevent a marked elevation of IOP in men undergoing RALP, a console time of <4 hours is important. Without a long console time, the use of RALP may be expanded to men having a high baseline IOP without compromising safety [37].

Based on the univariate linear regression during the CO_2 pneumoperitoneum in the sTp period, BMI, PAP, total CO_2 amount, and IAP were significant predictors of IOP changes, positively correlated with IOP. Multivariate analysis revealed that the only significant predictor of IOP was IAP. A possible explanation may be that a low IAP led to a decrease in the peritoneal CO_2 absorption and PAP, which may have attenuated the IOP increase [34]. Contrary, despite the fact that MAP, $EtCO_2$, and airway pressure remained consistently at the same level throughout the RALP and IOP levels still increased. But, there was no relation between rise in IOP and change in MAP, airway pressure, and $EtCO_2$ during surgery. Thus, this rise in IOP may be due to the obstruction in aqueous outflow in sTp with normal production and absorption of aqueous fluid. As the normalization of IOP occurs after normalization of patient posture, it can only be explained by sudden release of outflow obstruction of fluid leading to fall in pressure [36].

Regarding the contribution of preexisting retinal and/or central nervous system comorbidities on the risk of ophthalmic complications s following RALP with sTp, outcomes of patients with previous retinal surgery, cerebrovascular events, aneurysms, neurosurgery, and externally healthy comparators were compared—no retinal or CNS-related perioperative complications were reported in either the study or control groups [45]. In this study, IOP was not routinely recorded during the pre-, post-, or perioperative periods and, as such, it is unclear as to whether changes in IOP contributed to the incidence of retinal and CNS-related complications.

4. Anesthetic concerns for RALP

With the advent of robotic surgery, the physiologic change after pneumoperitoneum and sTp affect IOP. Use of the da Vinci surgical system requires additional precautions not normally needed for other laparoscopic procedures. For the anesthesiologist, robotic surgery comes with the following challenges: sTp to provide the best field of view for the surgeon, longer duration of pneumoperitoneum, and limited access to the patient after robot docking [46, 47]. This combination affects ocular, homeostasis, and leads to devastating ION. The anesthetic management of RALP patients involves a careful positioning on the operative table and appropriate fluid management [48]. RALP presents a challenge for anesthesiologist due to potentially serious inherent complications because of sTp and pneumoperitoneum during the procedure, therefore, anesthesiologists need to be fully aware of, prepared to handle, the challenges generated by this new technology, and manage the associated complications [49–55].

RALP has the following principal downside: a steep learning curve, although acceptable operative times can be achieved in <20 cases, and positive surgical margin rates may require experience with >80 cases before a plateau is achieved. The significant learning curve should not be understated [56]. A highly experienced surgeon can perform a RALP in about 105 minutes with a minimum blood loss of 111 ml [22]. But, every urologist needs to perform more than 150 operations to learn how to manipulate unfamiliar instruments and get used to a limited vision field. Until the operation is fully mastered by the surgeon, the time of surgical work can be multiplied, and blood loss will increase. As a result, complications may occur more frequently [57]. These data demonstrate the lowest IOP value in the supine position and the highest in sTp, regardless of the type of anesthesia. Almost all studies were conducted at 30-degree up to 35-degree sTp. The surgeon's working conditions are better, the steeper the positioning, the better the intra-abdominal view, and probably less bleeding. Hypotheses from previous studies show that patients placed in sTp for several hours have a high risk of ocular changes and perioperative complications.

IOP increases significantly between abdominal insufflation in supine position and 240 minutes of sTp. The greatest increase in IOP occurs within 5 minutes of placing the patient into the sTp and continues to increase significantly, while the patient is in sTp. IOP increases of the magnitude found in systematic review and meta-analysis demonstrate the need for implementing intraoperative interventions to mitigate the increase in IOP and reduce the risk for postoperative vision loss and other ocular complications in patients undergoing surgery in the sTp [58].

4.1 Anesthetic regimens during RALP

Anesthesia for patients undergoing robotic-assisted surgery is different from anesthesia for patients undergoing open or laparoscopic surgery, and new anesthetic concerns accompany robotic surgery. These concerns include physiological effects of fixed extreme positioning of the patient over a long time, pneumoperitoneum, and the need for carefully monitored relaxation of the patient. There is interest for prospective studies assessing perioperative ocular and visual events that add knowledge to the perioperative behavior of IOP in different types of surgery. Inhalation anesthesia with both sevoflurane and desflurane, as well as total intravenous anesthesia with propofol, are widely used in laparoscopic and robot-assisted procedures [59, 60]. There are no restrictions in the choice of the anesthetics [61].

Earlier, a systematic review with meta-analysis of 20 randomized controlled trials (RCT) comparing the effects of propofol-based total intravenous anesthesia (TIVA) and volatile anesthesia on IOP during laparoscopic surgery, including lower abdominal, colorectal, RALP (one study), cholecystectomy, pelvic and gynecological surgery, ophthalmic surgery, and spine surgery in the prone position was conducted. The mean IOP was significantly lower in the TIVA group after intubation, pneumoperitoneum, sTp, and lateral decubitus positioning. So, propofol-based TIVA is more effective during surgery at attenuating the elevation of IOP and should be considered in at-risk patients [62].

We found 14 full-text articles describing the effect of sTp on IOP during RALP under volatile or intravenous anesthesia. The included articles consisted of 13 prospective observational studies [23, 26, 31, 32, 35, 36, 40, 44, 63–68] and one clinical study (**Table 2**) [68]. All studies were prospective single-centered. As an intervention, there was a RCT of the effect of volatile versus intravenous anesthesia. The review did not include conference abstract because of the insufficient information for assessment of the quality of evidence [69], one article in Japanese [70], and one study with only graphically presented IOP levels [38].

First author, year		Awad 2009	Hoshikawa 2014	Yoo 2014		Mondzelewski 2015	Taketani 2015	Blecha 2017	Demasi 2017	Mizumoto 2017	Hirooka 2018	Hirooka 2019	Awad 2020	Goel 2020	Balkan 2021	Kondo 2021
Country		USA	Japan	South Korea		USA	Japan	Germany	Italy	Japan	Japan	Japan	USA	India	Turkey	Japan
Study design		Observational	Observational	RCT		Cohort	Observational	Observational	Observational	Observational	Observational	Observational	Cohort	Observational	Cohort	Observational
Patients, n		33	31	66		18	25	51	50	22	20	10	24	31	34	21
Anesthetic		Desflurane	Sevoflurane	Propofol	Sevoflurane	Sevoflurane	Propofol/ Desflurane	Propofol	Propofol	Propofol	Propofol/ Desflurane	Propofol/ Desflurane	NA	Propofol/ Sevoflurane	Sevoflurane	Sevoflurane
Total operative time		142 (105–210) min	4.57±0.03 h	92 (71–142) min	105 (85–140) min	NA	318±60 min	218 (120–357) min	143.56±7.98	5.46 h±60.3 min	274.4±52.2 min	227.3±47.9 min	248.2±40.2	150±41.7 min	205.3±44.3	263±61
sTp angle, °		25	23	30		30	25–30	45	30	30	30	30	25	>45	35–45	30
Pneumoperitoneum pressure, mmHg		15	NA	15±5		15	NA	15	15	NA	NA	NA	NA	15	10	NA
Estimated blood loss, mL		80 (45–155)	364±196	415±211	431±269	291±220	267±188	NA	213.2±98.6	350±343	173.3±158.9	90±131.7	134.2±115.6	NA	NA	NA
Intravenous fluid, mL		2000 (1600–3100)	NA	1485±588	1558±557	NA	1733±534	880 (550–2200)	1545±174	NA	NA	NA	1939.6±772.2	NA	NA	2230±460
IOP, mmHg	T_0	19.9±3.6[1]	13.2 (8–20)[2]	17.9±3.7[1]	17.5±3.6[1]	13.7±3.2[1]	14.9±2.1[1]	19.9±3.6[1]	19.9±3.3[1]	NA	NA	NA	NA	NA	NA	18.3±2.4[1]
	T_{min}	15.9±4.8[1]	9.8 (4–15)[2]	NA	NA	NA	11.0±2.7[1]	15.9±4.8[1]	13.1±3.5[1]	10.4 (8.2–12.5)[3]	12.3±2.6	11.2±3.8[1]	14.2 (11.7–16.6)[3]	19.2[1]	12.4±3.1[1]	NA
	T_{max}	33.9±7.4[1]*	24.2 (12–33)[2]*	19.9±3.8[1]	23.5±4.3[1]*	29,9 (27.4–32.5)[3]*#	29.4±7.0[1]*	33.9±7.4[1]*	32.2±4.4*	29.6 (27.6–31.5)[3]*	29.8±8.7*	28.3±4.8[1]*	37.4 (35.0–39.9)[3]*	40.0[1]*	21.8±4.7[1]*	25.3±2.2[1]*
		Ligation of the dorsal venous complex (SP)	End of sTp	30 min of sTp		60 min of surgery	4 h of sTp	Ligation of the SP	End of surgery	End of sTp	240 min of surgery	180 min of sTp	End of sTp	End of sTp	120 min of sTp	150 min of sTp
Ocular complications		Conjunctival edema in 7 patients; resolved the next day	None	None		Optic disc hemorrhage in 1 patient	Visual field defects in 7 patients 1 week after operation	None	None	None	None	None	None	None	None	None

Note: NA, not available; SP, Santorini's plexus.
The data of IOP are presented as [1]mean (±SD), [2]mean (range), [3]mean (95% CI), [4]median (IQR).
Time points for perioperative IOP measurements: T_0 - prior to induction of anesthesia while supine and awake, T_{min} - anesthetized and supine, and T_{max} –maximum during the operation in a sTp stay.
*Statistically significant differences: *between $T_{0/min}$ and T_{max}; #between patient's group.*

Table 2.
Comparative analysis of studies of IOP during RALP depending on the type of anesthesia.

The IOP value was 13.3 ± 0.58 mmHg higher on average at the end of the period of sTp compared with the supine position. One patient maintained significantly increased pressure and represented an outlier. However, adverse ophthalmic consequences of intraoperative changes in IOP were not identified [31].

The mean IOP increased three-fold time-dependently from 9.8 mmHg to 24.2 mmHg and a maximum of 36 mmHg after 4 hours of surgery. Not any statistically significant ocular complications related to IOP increase (changes of the retinal nerve fiber layer thickness or visual acuity) were observed [32].

Based on the results of RCT, it was reported that propofol-based TIVA was more effective than sevoflurane-based inhalational anesthesia in attenuating IOP increase during RALP with pneumoperitoneum and sTp. IOP was significantly less immediately after establishing pneumoperitoneum, 30 minutes after sTp, and 5 minutes after tracheal extubation in the operating room [68].

Significant elevations of IOP were experienced during robotic surgery utilizing sTp in patients with healthy eyes, operated under inhalation anesthesia of 17 up to 53 mmHg at 60 minutes of surgery, 24 up to 52 mmHg, and 24 to 55 mmHg at 150 and 240 minutes, respectively. There were no significant changes in retinal nerve fiber layer thickness and Humphrey visual field pattern [40].

Transient but significant unilateral visual field defects were found in 28% of the patients after RALP dominantly in the lower hemifield without abnormal findings in the optic nerve head or retina, and the visual field recovered to normal within 3 months after surgery. IOP was significantly increased up to 29.4 mmHg but did not differ significantly in patients anesthetized by inhalationally or intravenously [23].

The combination of permanent 45°sTp and pneumoperitoneum during RALP has a pronounced influence on IOP. The mean IOP was significantly 4 mmHg lower after induction of TIVA and more than doubled from 15.9 up to 33.9 mmHg at the end of the operation. In 14% of patients, IOP was higher than 40 mmHg, and the highest IOP measured was 59.6 mmHg. Non-ocular complications were observed in the recovery room, 8 hours later on the ward, and the next day [26].

sTp and pneumoperitoneum increased IOP from the induction of intravenous anesthesia directly proportional to the angle of tilt to 20 mmHg, so the pulsatile ocular blood flow and OPP both significantly decreased (from 15.5 ± 3.3 to 10.0 ± 3.2 μL/s and from 70.1 ± 5.9 to 51.7 ± 11.5 mmHg, respectively) and reached their lowest levels at the end of the RALP procedure. No complications occurred during the RALP procedure, nor were any postoperative complications reported [35].

IOP transiently significantly increased from 10.4 mmHg up to 29.6 mmHg with the timing of sTp in patients anesthetized intravenously. No significant disorders in ocular structural and functional parameters were found until long after RALP. The mean visual acuity, IOP, mean deviation, pattern standard deviation, the ganglion cell complex, and retinal nerve fiber layer thicknesses and the central fovea thicknesses measured before and after surgery did not differ significantly [44].

Although IOP significantly increased during RALP from 20 to 53 mmHg in healthy patients, regardless of the type of anesthesia, there was no progression of the visual field and retinal nerve fiber layer thicknesses after surgery or any other ocular complications at 1 and 3 months after surgery [63].

The effect of the sTp surgical procedure on the retinal structure and function during RALP in 10 glaucoma patients was investigated. Average IOP (mmHg) significantly increased from 11.2 ± 3.8 up to 28.3 ± 4.8 during RALP. Two eyes of two patients exhibited significant retinal nerve fiber layer thickness progression [64].

A significant increase in IOP during RALP secondary to the sTp and CO_2 insufflation was confirmed. Accumulation of subclinical damage on the retina, such as age, hypertension, diabetes, macular degeneration, anatomic breech position of optic disk at risk, and preexisting undetected ocular disease, could potentially lead to permanent changes when combined with sTp but not lead to significant retinal changes (ganglion cell complex and retinal nerve fiber layer thickness, foveal threshold, mean deviation, and pattern standard deviation) in patients with healthy ocular system at 3 months postoperative [65].

The study of the critical sTp angulation of patient with the floor more than 45° have shown the gradual rise in IOP with time. All the patients had a normal baseline mean IOP of 18–19 mmHg, remaining unchanged after creation of pneumoperitoneum indicating insignificant effect on IOP by raised IAP. As soon as the patient was placed in sTp, IOP started to rise with significant difference between IOP values at different time points with each other and baseline. The continuous rise in IOP reached maximum value of 40–41 mmHg. This information is important, especially in patients suffering from glaucoma and ocular hypertension as it may convert an advanced surgical technique into an ocular nightmare for the patient along with medicolegal issues for treating physician [36].

Changes in both IOP and optic nerve sheath diameter (ONSD) during RALP were evaluated, and any correlation between IOP and ONSD was examined during RALP. The highest IOP recorded in cohort was 32 mmHg and was associated with the 35-to 45-degree sTp and abdominal carbonic gas insufflation. Although both IOP and ONSD increased, no significant correlation between the two parameters was observed. This could be explained by the different mechanisms that increase IOP and ONSD. Based on these findings, intraoperative measurement of IOP can be useful in patients at high risk for increased IOP during RALP [66].

Small study assessed 21 patients to quantify changes in IOP in time in patients put on the sTp. In addition to confirming increased IOP during sTp, the authors concluded that the increase was moderate 90 minutes after positioning with a mean IOP value 7 mmHg above baseline, resuming to the baseline value as that before the induction of anesthesia roughly 30 minutes after returning to the supine position. No ocular complications, such as blindness, narrowing of the visual field, or impaired visual perception, were reported in the first 3 months postoperatively [67].

4.2 Other procedures

One study investigated whether IOP changes were different depending on the anesthetic drugs. Patients scheduled for pelvic laparoscopy were randomly allocated into the propofol-based TIVA group or the desflurane-based volatile group. In all the groups, IOP decreased after anesthesia was initiated (17 ± 2 to 11 ± 2 mmHg). Pneumoperitoneum in addition to the head-down position raised the IOP highly in the desflurane group, and the average IOP value was over the normal limit (22 ± 4 mmHg). In contrast, propofol kept IOP similar to the preoperative level during the whole period of pneumoperitoneum (18 ± 3 mmHg). For the laparoscopic surgery performed in the head-down position, propofol may be more helpful in preventing ocular hypertension [71].

The effects of propofol-based TIVA with those of sevoflurane anesthesia on IOP in patients undergoing lower abdominal laparoscopic surgery in sTp were compared. The change in IOP was significantly different between the groups. Maximum rise in IOP was 15.5 ± 0.9 mmHg and 19.8 ± 1.2 mmHg in TIVA group and sevoflurane group, respectively. In TIVA group, IOP remained almost equal to the baseline value, while in sevoflurane group, IOP increased significantly with the difference 4.0 ± 1.2 mmHg. So, TIVA is more effective than inhalational anesthesia in attenuating the increase in IOP during laparoscopic surgery requiring pneumoperitoneum and sTp [72].

IOP variation during repetitive positional changes in patients undergoing laparoscopic colorectal surgery was evaluated, and the effect of desflurane and propofol anesthesia on IOP change was compared. Repetitive positional changes in anesthetized patients caused markedly increased IOP in the sTp. IOP values were significantly lower in patients undergoing TIVA than in patients undergoing desflurane anesthesia during intraoperative positional changes. TIVA was more effective than inhalation anesthesia in attenuating IOP increases during frequent positional changes in long-duration laparoscopic surgeries. IOP values in sTp in the desflurane group were higher than those in the propofol group. In contrast, most of the IOP values in the sTp in the propofol group remained within the normal range [73].

Effects of combinations of four different anesthetic agents on IOP in laparoscopic gynecological operations using 12–14 mmHg pneumoperitoneum and sTp at 35° were studied in a prospective double-blind RCT. Different from other studies, two intravenous agents used in anesthesia induction were combined with two inhalation anesthetics, and their effects on IOP were investigated. Results indicate that propofol induction causes decreased changes in IOP, independent of sevoflurane, or desflurane use. In addition, there were no statistically significant differences between the IOPs in supine position after extubation and before intubation. Thus, the IOP values in supine position after extubation were not affected by the anesthesia technique. In all groups, rather than pneumoperitoneum, sTp increased IOP values by higher amounts [74].

4.3 Effects of dexmedetomidine

Whether dexmedetomidine effectively attenuates the increase in IOP remains inconclusive. The aim of systematic review and meta-analysis was to evaluate the effects of dexmedetomidine on IOP in adult patients undergoing surgery, which requires general anesthesia and endotracheal intubation. Twenty-nine RCTs were included. The IOP levels were significantly lower in patients receiving dexmedetomidine after the administration, after pneumoperitoneum, and after the patients were placed in a sTp. So, dexmedetomidine effectively attenuates the increase in IOP levels and should be considered, especially for at-risk patients [75].

Regarding the effect of dexmedetomidine, a RCT evaluated the effect of intraoperative continuous infusion versus equal volume of physiologic saline on IOP in patients undergoing RALP in the sTp. The highest mean IOP measured 60 minutes after the patients had been placed in the sTp, was 19.9 ± 5.0 mmHg in dexmedetomidine group, and 25.7 ± 5.0 mmHg in control group with significant difference. No ocular complications were noted. So, intraoperative continuous infusion of dexmedetomidine may help alleviate IOP increase in patients undergoing RALP in the sTp [76].

Also, a prospective double-blinded RCT assessed the efficacy of systemically infused dexmedetomidine in preventing the increase in IOP caused by a sTp. Patients undergoing laparoscopic or robotic-assisted surgery due to colon, prostate, or gynecological cancer received dexmedetomidine throughout the operation versus saline. IOP increased in the sTp and was 11.3 mmHg higher at the end of surgery in the saline group. This increase in IOP was attenuated in the dexmedetomidine group, for which

IOP was only 4.2 mmHg higher. So, dexmedetomidine infusion attenuated the increase in IOP during laparoscopic surgery in a sTp, without further decreasing the OPP [77].

The effect of intraoperative dexmedetomidine on the IOP in patients undergoing RALP under propofol-remifentanil anesthesia was studied in a double-blind RCT. A linear mixed model analysis demonstrated IOP at 180 minutes after placing patient in the sTp, significantly lower in the dexmedetomidine group than in the control group. So, dexmedetomidine combined with propofol decreases IOP in the sTp during RALP [78].

4.4 Other ways to reduce the elevated IOP

While the literature is infrequent and undeveloped, certain anesthetic techniques, including deep neuromuscular blockade, modified positioning, providing periodic position changes or rest periods, and administering specific medications or anesthesia technique have been shown to mild-to-modest attenuate the increase in IOP [79].

One of the ways to reduce IOP during RALP may be to change the extreme head-down position, and in an RCT, modified Z Trendelenburg position was used (the patient's head and shoulders are placed horizontally). Median IOP was in the normal range at anesthesia induction and before positioning and increased at sTp. From the start of modified Z Trendelenburg position, IOP decreased and was significantly lower. At the time of supine and the end of pneumoperitoneum, IOP decreased to normal (19.6 mmHg) in Z Trendelenburg position group but remained in the hypertensive range (24.9 mmHg) in sTp group. The significant positive effect on patient neuro-ocular safety by lowering IOP and accelerating its recovery to the normal range was reached without any negative consequences for surgery [80].

Contrary, increasing the angle of the operating table from 25-degree to 30-degree would provide better surgical visibility, which would lead to shorter surgery time and reduced blood loss in RALP. In a prospective RCT involved a total of 30 consecutive patients, significant time-dependent increases in IOP were observed in both the 25-degree and 30-degree Trendelenburg positions; however, the IOP values measured at the same time points were similar between the two groups. Several operative variables (DoS, EBL, and intravenous fluid intake) did not significantly differ between the two groups. So, the 25-degree sTp can reduce the risk of catastrophic position-related ophthalmologic complications after RALP without prolonging the operative time and/or increasing EBL during surgery, as compared with the 30-degree sTp [81].

Upon identification of elevated IOP readings intraoperatively, we can effect changes in position. In some patients with severe IOP elevation during sTp, when the head of the operating table was returned to the level, supine position halfway through the procedure, to meet procedural requirements, IOP appeared to improve by the end of the procedure in contrast to subjects who were not leveled. Observations that elevation of IOP over time can be mitigated by transient and periodic changes in position during surgery, open the question of whether interventions other than maintenance of MAP can decrease the effect of elevated IOP on cerebral perfusion pressure. It is possible that by measuring IOP, one can predict when a change in the level of the table should be made so as to limit further IOP increases. A second intervention is that, because the calculation of perfusion and flow to the eye (OPP) is derived by obtaining IOP readings and subtracting them from the MAP, one can maintain OPP by elevating MAP. In summary, by measuring IOP, practice changes can be implemented to ensure the patient's ophthalmic safety [82].

There has been no study of the effect of positive end-expiratory pressure (PEEP) on IOP during pneumoperitoneum with sTp. Applying 5 cmH_2O of PEEP as compared with zero-PEEP did not significantly increase IOP during RALP. These results suggest that low PEEP <10 cm H_2O can be safely used during RALP that takes a few hours of pneumoperitoneum and sTp without a clinically significant risk of IOP increase in patients without preexisting eye disease [83].

The possible effect of perioperative fluid management on the outcome of surgical patients has recently been debated. One of the inherent risks of the liberal approach to infusion therapy is unintentional hypervolemia, which can lead to an increase in IOP. Restrictive compared to liberal intravenous fluid administration leads to a better patient's outcomes and reduction in the length of hospital stay. In a small group of gynecological patients during robotic surgery, restrictive strategy, along with maintaining close to normal $EtCO_2$ levels, negated the effects of sTp, and increased PAP on IOP [84]. Contrary, in another small group of women undergoing laparoscopic gynecologic pelvic surgery, the effect of liberal versus restrictive protocols of perioperative fluid management on fluctuations in IOP during the perioperative period was explored. The main finding of this prospective study was the similar IOP measurements between patients who were treated with different preoperative fluid management protocols [85].

Continuous deep neuromuscular blockage (NMB) can improve surgical conditions and facilitate RALP to significantly attenuate the increase in IOP, as was shown in the double-blind RCT. The highest IOP value was observed at 60 minutes after CO_2 pneumoperitoneum in the sTp and was significantly lower in deep NMB group (19.8 ± 2.1 mmHg) than in the moderate NMB group (23.3 ± 2.7 mmHg). RALP was accomplished at the low IAP of 8 mmHg in 25 and 88%, respectively. The overall surgical condition was acceptable in both groups [34].

The effects of continuous systemic administration of esmolol versus placebo on IOP during laparoscopic and robotic surgeries for recto-sigmoid cancer in a sTp were investigated. The IOP increased markedly after adopting the sTp, reaching 28.8 ± 4.4 mmHg, which was ~5.7 mmHg higher than in the esmolol group. So, esmolol can alleviate the increase in IOP during a sustained sTp without adverse effects [86].

A dangerous increase in IOP during prolonged laparoscopic intervention may be susceptible to topical drugs. In a quasi-experimental study, dorzolamide-timolol eye drops were examined during lengthy laparoscopic urologic and gynecologic procedures with the patient in sTp. Patients received treatment when IOP levels reached 38 to 40 mmHg. Repeated-measures analysis of variance showed that IOP values dropped significantly after drug intervention at 60-, 90-, and 120 minutes. Effect size of pharmacologic intervention on IOP reduction was strong [33]. On the contrary, in prospective masked interventional RCT, the effect of preoperative administration of brimonidine tartrate versus placebo (artificial tears) on IOP during RALP in sTp setting was evaluated. Significant and sustained IOP elevation of >1.5x baseline in the sTp was noted in both groups. The mean IOP 1 hour after sTp was 29.4 ± 6.9 and 27.2 ± 3.4 mmHg in the drug and placebo groups, respectively (P = 0.35). So, preoperative brimonidine does not prevent IOP spikes in sTp [87].

5. IOP in patients with glaucoma

Although acute angle closure glaucoma is increasingly prevalent and often questions arise concerning perioperative anesthetic management, evidence-based recommendations to guide safe anesthesia care in patients with glaucoma are currently lacking. Patients with low vision present challenges to the anesthesia provider that are becoming more common as the population ages [88].

Patients with primary open-angle glaucoma have decreased outflow through the trabecular meshwork of the eye, resulting in IOP. It is known that the sTp causes increased IOP, but there are no current guidelines for monitoring and treating patients with glaucoma undergoing surgical procedures while in the sTp. A case of successful intraoperative management of increased IOP in a patient with glaucoma undergoing RALP, while in sTp was described [89].

Prospective study enrolled 39 patients undergoing laparoscopic surgery, including 10 with eye diseases (six with normal tension glaucoma and four with a narrow anterior chamber and normal range IOP). All patients, with or without eye diseases, experienced significantly elevated IOP with no significant differences between groups. The average maximal IOP reached 20 mmHg at the end of surgery, with no patient had an IOP of >40 mmHg as a critical threshold during surgery. So, using pneumoperitoneum of 10 mmHg and a Trendelenburg position of 15° during a 3-hour surgical period could be performed within a safe range of IOP [90].

Given the aging population of RALP patients in whom risk for glaucoma is significant, preoperative ocular health assessment should be considered. Robotic technology for prostatectomy is increasingly frequent. Such procedures, as for robotic surgery for other pelvic procedures, require steep head-down tilt body positioning, with transient increase in IOP, and it is a relative counterindication for patients with open-angle glaucoma, aimed at avoiding additional damage to the optical nerve [31]. Specific attention should be directed toward identifying any history of glaucoma during the preoperative evaluation of patients in whom RALP is planned. If the patient has a known diagnosis of glaucoma, a discussion of the perioperative management with the ophthalmologist is recommended [91]. Since an increased IOP during the surgery was the probable cause of the retinal nerve fiber layer thickness changes, ophthalmologic examinations should be performed before and after RALP, especially in glaucoma patients. While it remains unknown whether eyedrops can effectively reduce elevated IOP during surgery, glaucoma patients should be instructed to take their medications prior to RALP when using the sTp and have their IOP routinely monitored during these surgeries. Furthermore, in patients scheduled to undergo RALP, comprehensive eye examinations should be recommended in all these subjects prior to the surgical procedure [64].

6. Recommendations for robotic-assisted and laparoscopic procedures

Most of the complications assessed are related specifically to RALP approaches, and therefore are not expected in open surgery. The present review is focused on avoidable RALP-related ophthalmic complications and suggests how the surgical team should work in a way to avoid them. In the light of evidence from the existing literature, surgical and anesthesiologic measures to prevent and manage ocular complications in robotic-assisted laparoscopic interventions are suggesting. It is advisable to develop an interdisciplinary collaboration between surgeons, anesthesiologists, and ophthalmologists on a procedural and medicolegal level with the intent of mutual training [92]. Some advocate preoperative ocular examination and regular monitoring of ONSD as a representative of raised intracranial and intraocular pressure during intraoperative period to ensure early awareness of surgical team, implement early interventions as needed, and thus reduce intraoperative ocular complications and POVL.

A Molloy/Bridgeport Anesthesia Associates Observation Scale enabling caregivers to determine when to institute preventive measures to optimize ocular perfusion was designed based on a prospective repeated-measures correlation regression model. Visual assessment of presence of eyelid edema or chemosis and baseline IOP values determine the probability of when an IOP greater than 40 mmHg (critical threshold) is reached. Significant predictors of IOP greater than 40 mmHg are determined to be presence of chemosis and baseline IOP and significantly correlated to increasing IOP. The receiver operating characteristic curve area under the curve score is 0.86 ± 0.03. Caregivers can use this observation scale to assess the need and timing for IOP-normalizing interventions and possibly to prevent POVL [82].

On 11 September 2012, the APSF convened a multidisciplinary consensus of experts to create a statement of safety recommendations for those patients at-risk, and to assure that suggested management reflects evolving information. Those experts explained how (1) sTp in robotic pelvic surgery, (2) IAP from insufflation, (3) low colloid/crystalloid ratios, and (4) a long duration of such conditions could all increase the risk of retro-orbital and/or optic nerve edema and compartment syndrome. These hydrostatic and osmotic forces are the variable components within the Starling equation that describe net fluid movement between compartments [93].

Joint consensus on anesthesia in urologic and gynecologic robotic surgery, written by Società Italiana di Anestesia Analgesia Rianimazione e Terapia Intensiva (SIAARTI), Società Italiana di Ginecologia e Ostetricia (SIGO), and Società Italiana di Urologia (SIU), recommend to limit the use of 30 degrees sTp only for the time strictly necessary for surgery to be performed, using a position tailored to the pelvic operatory field of the subject; the sTp should be avoided in high risk-patients (Level moderate, Grade A) [94].

Whether patients should be informed of ION, especially those undergoing higher-risk surgery and complex instrumented surgery, is controversial. Because most patients are first seen by anesthesiologists soon before surgery, consider requesting that the surgeon discuss the possible complication at an earlier, more relaxed, preoperative visit [95].

Under the auspices of the Anesthesia Patient Safety Foundation, an interdisciplinary consensus of experts has been created to make recommendations on patient safety [93]. The experts explained that (1) abdominal position, (2) sTp, (3) intra-abdominal pressure, (4) colloid-to-crystalloid ratio, and (5) prolonged duration of such conditions increase the risk of developing optic nerve edema and compartment syndrome [96].

Given the lack of scientific proof of cause and effect, the informed consent process itself was emphasized. The APSF thinks that anesthesia professionals and surgeons should discuss, with those patients at-risk, the remote chance of partial or complete blindness, the current state of understanding of those risks, and the interventions that may reduce those risks. Furthermore, the APSF considers that if this is not in a joint consent, nor part of the surgical consent, then it should be made a part of the anesthesia consent. Suggested and speculative interventions might include the following: minimizing intra-abdominal pressure, degree and duration of Trendelenburg, and the amount of crystalloid; deliberate hypotension should be carefully reconsidered, the head may be elevated, colloid may be substituted for some of the crystalloid, anemia should be monitored, and considerations of staging the procedure could be made [93].

7. Conclusions

Robotic surgery is becoming more prevalent and is being increasingly used in various specialties. RALP presents a challenge not only for surgeons but also for anesthesiologists. The detailed understanding of physiological changes of RALP is essential. The performing of the RALP requires the use of sTp. With the increasing popularity, frequency, and acceptance of the RALP procedure, an awareness of unique intra- and postoperative complications is heightened, including that of increases in IOP. The sTp required for operative exposure has been shown to increase this value. We found that IOP increases significantly depending on the duration of the operation. Despite this increase, there were small or no significant postoperative changes in visual function and ophthalmic complications in patients without prior eye disease.

There are several perioperative factors involved in the increase in IOP. Some of these factors, such as hemodynamic management, ventilation strategy, and volemic load, can be controlled by an anesthesiologist. Other factors, such as patient positioning and duration, are inherent in the operation itself. The exact nature of the relationship between the angle and duration of sTp and the increase in IOP remains unclear.

Most patients generally tolerate robotic prostatectomy well and appreciate the benefits; however, anesthesiologists must have an intimate knowledge of the physiological changes associated with RALP. Specifically, anesthesiologists must consider the changes in the cardiopulmonary, ocular, and intracranial systems that occur when patients are placed in the lithotomy and sTp, and when pneumoperitoneum is created.

Intraoperative head-low position had certain effects on the IOP of patients, showing different patterns of changes with the surgical process. Therefore, the head-low angle should be minimized without affecting the surgical operation. Intraoperative IOP measurement is recommended for patients in head-low foot-high lithotomy position. After the operation, the nurses should make a good handover and pay attention to the postoperative IOP changes and the occurrence of any adverse events, to identify the problems early and provide appropriate treatment.

While the literature is infrequent and undeveloped, certain anesthetic parameters, including deep neuromuscular blockade, modified positioning, and the use of dexmedetomidine have been shown to have mild-to-modest decreases in IOP. These modifications may prove to have even greater significance in patients with preexisting ophthalmologic pathologies, such as glaucoma, which were excluded from the studies' analyses. Well-selected patients, adequate positioning, mentorship training during the learning curve, and avoiding last-longing procedures are key steps to prevent RALP-related complications. Fortunately, those specific complications are rare, but one should keep alert as they can be devastating if not recognized early, thus surgeons should have a low threshold of suspicion. A dedicated robotic team is essential to reduce perioperative complications. Hence, anesthesiologists need to stay abreast of current knowledge and be prepared to give better quality of anesthesia care to patients. Meticulous preoperative ophthalmological assessment, restriction of intravenous fluids, "rest stops," eyelid taping, and ocular dressings are the major protective measures suggested by the literature. Collaboration between the surgical team and the anesthetist is also essential. Further studies should focus on resident training, cost-effectiveness, and long-term outcomes in anesthesia for robotic surgery.

Careful preoperative evaluation, intraoperative conduction, minimizes the risk of complications and helps patients to reach full recovery. Excellent outcomes are the result of individualized approach to the patient and good communication between team members. Large prospective studies are needed to assess the relationship between sTp and ophthalmic complications and to develop clinical recommendations for the prevention and treatment of elevated IOP in elderly patients with preexisting eye diseases during operations of longer duration. Up to this point, it is necessary to consider the available data of the effect of patient positioning on IOP. Aggressive assessment of elevated IOP may be required when the operation time exceeds 5 hours. A patient with glaucoma may not be a candidate for robotic surgery due to the risk of ophthalmic complications.

Author details

Ildar Lutfarakhmanov*†, Alyona Lifanova†, Peter Mironov† and Valentine Pavlov†
Bashkir State Medical University, Ufa, Russia

*Address all correspondence to: lutfarakhmanov@yandex.ru

† These authors contributed equally.

References

[1] George EI, Brand TC, LaPorta A, Marescaux J, Satava RM. Origins of robotic surgery: From skepticism to standard of care. JSLS. 2018;**22**(4): e2018.00039. DOI: 10.4293/JSLS.2018.00039

[2] Sheetz KH, Claflin J, Dimick JB. Trends in the adoption of robotic surgery for common surgical procedures. JAMA Network Open. 2020;**3**(1):e1918911. DOI: 10.1001/jamanetworkopen.2019.18911

[3] Gkegkes ID, Karydis A, Tyritzis SI, Iavazzo C. Ocular complications in robotic surgery. The International Journal of Medical Robotics. 2015;**11**(3): 269-274. DOI: 10.1002/rcs.1632

[4] Lee LA. Visual loss, venous congestion and robotic prostatectomies. ASA Newsletter. 2011;**75**:26-27

[5] Curtillet E. Postoperative amaurosis. Afrique Française Chirurgicale. 1950;**1-3**: 51-53

[6] Hewer CL. The physiology and complications of the Trendelenburg position. Canadian Medical Association Journal. 1956;**74**(4):285-288

[7] Lee LA, Newman N. Postoperative visual loss after anesthesia for nonocular surgery. UpToDate. 2018. Available from: https://www.uptodate.com/contents/postoperative-visual-loss-after-anesthesia-for-nonocular-surgery [Accessed: December 24, 2018]

[8] Foroozan R. Shock-induced anterior ischaemic optic neuropathy after radical prostatectomy. Clinical & Experimental Ophthalmology. 2004;**32**(4):438-439. DOI: 10.1111/j.1442-9071.2004.00851.x

[9] Berg KT, Harrison AR, Lee MS. Perioperative visual loss in ocular and nonocular surgery. Clinical Ophthalmology. 2010;**4**:531-546. DOI: 10.2147/opth.s9262

[10] Todd MM. Good news: But why is the incidence of postoperative ischemic optic neuropathy falling? Anesthesiology. 2016;**125**(3):445-448. DOI: 10.1097/ALN.0000000000001212

[11] Kan KM, Brown SE, Gainsburg DM. Ocular complications in robotic-assisted prostatectomy: A review of pathophysiology and prevention. Minerva Anestesiologica. 2015;**81**(5): 557-566

[12] Kla KM, Lee LA. Perioperative visual loss. Best Practice & Research. Clinical Anaesthesiology. 2016;**30**(1):69-77. DOI: 10.1016/j.bpa.2015.11.004

[13] Molloy BL. Implications for postoperative visual loss: Steep Trendelenburg position and effects on intraocular pressure. AANA Journal. 2011;**79**(2):115-121

[14] Alwon K, Hewer I. Perioperative vision loss: Considerations and management. AANA Journal. 2016; **84**(5):363-370

[15] Goepfert CE, Ifune C, Tempelhoff R. Ischemic optic neuropathy: Are we any further? Current Opinion in Anaesthesiology. 2010;**23**(5):582-587. DOI: 10.1097/ACO.0b013e32833e15d0

[16] Newman NJ. Perioperative visual loss after nonocular surgeries. American Journal of Ophthalmology. 2008;**145**(4): 604-610. DOI: 10.1016/j.ajo.2007.09.016

[17] Kelly DJ, Farrell SM. Physiology and role of intraocular pressure in contemporary anesthesia. Anesthesia

and Analgesia. 2018;**126**(5):1551-1562. DOI: 10.1213/ANE.0000000000002544

[18] Freshcoln M, Diehl MR. Repositioning during robotic procedures to prevent postoperative visual loss. OR Nurse. 2014;**8**(4):36-41

[19] Ghomi A. Robotics in practice: New angles on safer positioning. Contemporary OB/GYN. 2012;**57**(10):26-37

[20] Weber ED, Colyer MH, Lesser RL, Subramanian PS. Posterior ischemic optic neuropathy after minimally invasive prostatectomy. Journal of Neuro-Ophthalmology. 2007;**27**(4): 285-287. DOI: 10.1097/WNO.0b013e31815b9f67

[21] Mizrahi H, Hugkulstone CE, Vyakarnam P, Parker MC. Bilateral ischaemic optic neuropathy following laparoscopic proctocolectomy: A case report. Annals of the Royal College of Surgeons of England. 2011;**93**(5):e53-e54. DOI: 10.1308/147870811X582828

[22] Patel VR, Palmer KJ, Coughlin G, Samavedi S. Robot-assisted laparoscopic radical prostatectomy: Perioperative outcomes of 1500 cases. Journal of Endourology. 2008;**22**(10):2299-2305. DOI: 10.1089/end.2008.9711

[23] Taketani Y, Mayama C, Suzuki N, Wada A, Oka T, Inamochi K, et al. Transient but significant visual field defects after robot-assisted laparoscopic radical prostatectomy in deep Trendelenburg position. PLoS One. 2015; **10**(4):e0123361. DOI: 10.1371/journal.pone.0123361

[24] Ozcan AA, Ulas B. Ischemic optic neuropathy in robotic-assisted gynaecologic surgery: A case report. The Journal of Obstetrics and Gynaecology Research. 2019;**45**(3):748-750. DOI: 10.1111/jog.13877

[25] Radtke S, Florence E, Clavijo A, Do L, Lopez I. Complete vision loss after laparoscopic hysterectomy. Case Reports in Obstetrics and Gynecology. 2021; **2021**:6643703. DOI: 10.1155/2021/6643703

[26] Blecha S, Harth M, Schlachetzki F, Zeman F, Blecha C, Flora P, et al. Changes in intraocular pressure and optic nerve sheath diameter in patients undergoing robotic-assisted laparoscopic prostatectomy in steep 45° Trendelenburg position. BMC Anesthesiology. 2017;**17**(1):40. DOI: 10.1186/s12871-017-0333-3

[27] Bhoopat T, Chansaenroj P. Comparison of intraocular pressure during laparoscopic totally extraperitoneal (TEP) versus transabdominal preperitoneal (TAPP) inguinal hernia repair. Surgical Endoscopy. 2022;**36**(3):2018-2024. DOI: 10.1007/s00464-021-08487-x

[28] Borahay MA, Patel PR, Walsh TM, Tarnal V, Koutrouvelis A, Vizzeri G, et al. Intraocular pressure and steep Trendelenburg during minimally invasive gynecologic surgery: Is there a risk? Journal of Minimally Invasive Gynecology. 2013;**20**(6):819-824. DOI: 10.1016/j.jmig.2013.05.005

[29] Adisa AO, Onakpoya OH, Adenekan AT, Awe OO. Intraocular pressure changes with positioning during laparoscopy. JSLS. 2016;**20**(4): e2016.00078. DOI: 10.4293/JSLS.2016.00078

[30] Wayne G, Wei J, Atri E, Wong V, Garcia-Gil M, Pereira J, et al. Trends in positioning for robotic prostatectomy: Results from a survey of the endourological society. Cureus. 2021; **13**(1):e12628. DOI: 10.7759/cureus.12628

[31] Awad H, Santilli S, Ohr M, Roth A, Yan W, Fernandez S, et al. The effects of steep Trendelenburg positioning on intraocular pressure during robotic radical prostatectomy. Anesthesia and Analgesia. 2009;**109**(2):473-478. DOI: 10.1213/ane.0b013e3181a9098f

[32] Hoshikawa Y, Tsutsumi N, Ohkoshi K, Serizawa S, Hamada M, Inagaki K, et al. The effect of steep Trendelenburg positioning on intraocular pressure and visual function during robotic-assisted radical prostatectomy. The British Journal of Ophthalmology. 2014;**98**(3):305-308. DOI: 10.1136/bjophthalmol-2013-303536

[33] Molloy B, Cong X. Perioperative dorzolamide-timolol intervention for rising intraocular pressure during steep Trendelenburg positioned surgery. AANA Journal. 2014;**82**(3):203-211

[34] Yoo YC, Kim NY, Shin S, Choi YD, Hong JH, Kim CY, et al. The intraocular pressure under deep versus moderate neuromuscular blockade during Low-pressure robot assisted laparoscopic radical prostatectomy in a randomized trial. PLoS One. 2015;**10**(8):e0135412. DOI: 10.1371/journal.pone.0135412

[35] Demasi CL, Porpiglia F, Tempia A, D'Amelio S. Ocular blood flow in steep Trendelenburg positioning during robotic-assisted radical prostatectomy. European Journal of Ophthalmology. 2018;**28**(3):333-338. DOI: 10.5301/ejo.5001061

[36] Goel N, Chowdhury I, Dubey J, Mittal A, Pathak S. Quantitative rise in intraocular pressure in patients undergoing robotic surgery in steep Trendelenburg position: A prospective observational study. Journal of Anaesthesiology Clinical Pharmacology. 2020;**36**(4):546-551. DOI: 10.4103/joacp.JOACP_96_20

[37] Shirono Y, Takizawa I, Kasahara T, Maruyama R, Yamana K, Tanikawa T, et al. Intraoperative intraocular pressure changes during robot-assisted radical prostatectomy: Associations with perioperative and clinicopathological factors. BMC Urology. 2020;**20**(1):26. DOI: 10.1186/s12894-020-00595-5

[38] Chen K, Wang L, Wang Q, Liu X, Lu Y, Li Y, et al. Effects of pneumoperitoneum and steep Trendelenburg position on cerebral hemodynamics during robotic-assisted laparoscopic radical prostatectomy: A randomized controlled study. Medicine (Baltimore). 2019;**98**(21):e15794. DOI: 10.1097/MD.0000000000015794

[39] Sun Y, Wang J, Wang W, Fan G, Wu S, Zhao F, et al. Effect of different surgical positions on intraocular pressure: A cross-sectional study. BMC Ophthalmology. 2022;**22**(1):318. DOI: 10.1186/s12886-022-02547-z

[40] Mondzelewski TJ, Schmitz JW, Christman MS, Davis KD, Lujan E, L'Esperance JO, et al. Intraocular pressure during robotic-assisted laparoscopic procedures utilizing steep Trendelenburg positioning. Journal of Glaucoma. 2015;**24**(6):399-404. DOI: 10.1097/IJG.0000000000000302

[41] Pinkney TD, King AJ, Walter C, Wilson TR, Maxwell-Armstrong C, Acheson AG. Raised intraocular pressure (IOP) and perioperative visual loss in laparoscopic colorectal surgery: A catastrophe waiting to happen? A systematic review of evidence from other surgical specialities. Techniques in Coloproctology. 2012;**16**(5):331-335. DOI: 10.1007/s10151-012-0879-5

[42] Grosso A, Scozzari G, Bert F, Mabilia MA, Siliquini R, Morino M. Intraocular pressure variation during

colorectal laparoscopic surgery: Standard pneumoperitoneum leads to reversible elevation in intraocular pressure. Surgical Endoscopy. 2013;**27**(9): 3370-3376. DOI: 10.1007/s00464-013-2919-2

[43] Ozcan MF, Akbulut Z, Gurdal C, Tan S, Yildiz Y, Bayraktar S, et al. Does steep Trendelenburg positioning effect the ocular hemodynamics and intraocular pressure in patients undergoing robotic cystectomy and robotic prostatectomy? International Urology and Nephrology. 2017;**49**(1): 55-60. DOI: 10.1007/s11255-016-1449-y

[44] Mizumoto K, Gosho M, Iwaki M, Zako M. Ocular parameters before and after steep Trendelenburg positioning for robotic-assisted laparoscopic radical prostatectomy. Clinical Ophthalmology. 2017;**11**:1643-1650. DOI: 10.2147/OPTH.S139874

[45] Chalmers D, Cusano A, Haddock P, Staff I, Wagner J. Are preexisting retinal and central nervous system-related comorbidities risk factors for complications following robotic-assisted laparoscopic prostatectomy? International Brazilian Journal of Urology. 2015;**41**(4):661-668. DOI: 10.1590/S1677-5538.IBJU.2014.0464

[46] Lee JR. Anesthetic considerations for robotic surgery. Korean Journal of Anesthesiology. 2014;**66**(1):3-11. DOI: 10.4097/kjae.2014.66.1.3

[47] Tameze Y, Low YH. Outpatient robotic surgery: Considerations for the anesthesiologist. Advances in Anesthesia. 2022;**40**(1):15-32. DOI: 10.1016/j.aan.2022.06.001

[48] Hernandez-Meza G, Gainsburg DM. Anesthetic concerns for robotic-assisted laparoscopic radical prostatectomy: An update. Minerva Anestesiologica. 2023; **89**(9):812-823. DOI: 10.23736/S0375-9393.23.17284-1

[49] Awad H, Walker CM, Shaikh M, Dimitrova GT, Abaza R, O'Hara J. Anesthetic considerations for robotic prostatectomy: A review of the literature. Journal of Clinical Anesthesia. 2012;**24**(6):494-504. DOI: 10.1016/j.jclinane.2012.03.003

[50] Bačak Kocman I, Mihaljević S, Goluža E, Peršin Beraković A, Margaretić Piljek N, Kuliš T, et al. Anesthesia for robot-assisted radical prostatectomy - a challenge for anaesthesiologist. Acta Clinica Croatica. 2022;**61**(Suppl. 3):76-80. DOI: 10.20471/acc.2022.61.s3.11

[51] Chatti C, Corsia G, Yates DR, Vaessen C, Bitker MO, Coriat P, et al. Complications de l'anesthésie générale inhérentes à la voie laparoscopique et à la prostatectomie totale robot-assistée [prevention of complications of general anesthesia linked with laparoscopic access and with robot-assisted radical prostatectomy]. Progrès en Urologie. 2011;**21**(12):829-834. DOI: 10.1016/j.purol.2011.05.004

[52] Kakar PN, Das J, Roy PM, Pant V. Robotic invasion of operation theatre and associated anaesthetic issues: A review. Indian Journal of Anaesthesia. 2011;**55**(1):18-25. DOI: 10.4103/0019-5049.76577

[53] Kapur A, Kapur V. Robotic surgery: Anaesthesiologist's contemplation. Malaysian Journal of Medical Sciences. 2020;**27**(3):143-149. DOI: 10.21315/mjms2020.27.3.15

[54] Katayama S, Mori K, Pradere B, Yanagisawa T, Mostafaei H, Quhal F, et al. Influence of steep Trendelenburg position on postoperative complications:

A systematic review and meta-analysis. Journal of Robotic Surgery. 2022;**16**(6): 1233-1247. DOI: 10.1007/s11701-021-01361-x

[55] Suryawanshi CM, Shah B, Khanna S, Ghodki P, Bhati K, Ashok KV. Anaesthetic management of robot-assisted laparoscopic surgery. Indian Journal of Anaesthesia. 2023;**67**(1): 117-122. DOI: 10.4103/ija.ija_966_22

[56] Murphy DG, Bjartell A, Ficarra V, Graefen M, Haese A, Montironi R, et al. Downsides of robot-assisted laparoscopic radical prostatectomy: Limitations and complications. European Urology. 2010;**57**(5):735-746. DOI: 10.1016/j.eururo.2009.12.021

[57] Kitamura S, Takechi K, Yasuhira A, Fujioka S. Changes in intraocular pressure in patients with glaucoma during robotic-assisted laparoscopic radical prostatectomy. The Japan Society for Clinical Anesthesia. 2017;**37**(7): 743-747. DOI: 10.2199/jjsca.37.743

[58] Wicklin VAN, SA. Systematic review and meta-analysis of prone position on intraocular pressure in adults undergoing surgery. International Journal of Spine Surgery. 2020;**14**(2): 195-208. DOI: 10.14444/7029

[59] Danic MJ, Chow M, Alexander G, Bhandari A, Menon M, Brown M. Anesthesia considerations for robotic-assisted laparoscopic prostatectomy: A review of 1,500 cases. Journal of Robotic Surgery. 2007;**1**(2):119-123. DOI: 10.1007/s11701-007-0024-z

[60] Gainsburg DM. Anesthetic concerns for robotic-assisted laparoscopic radical prostatectomy. Minerva Anestesiologica. 2012;**78**(5):596-604

[61] Börgers A, Brunkhorst V, Groeben H. Anästhesie in der Urologie - Anästhesie bei roboterassistierter Prostatektomie [Anaesthesia for urological surgery - Anaesthesia for robotic assisted prostatectomies]. Anästhesiologie, Intensivmedizin, Notfallmedizin, Schmerztherapie. 2013; **48**(7–8):488-493. German. DOI: 10.1055/s-0033-1352496

[62] Chang CY, Chien YJ, Wu MY. Attenuation of increased intraocular pressure with propofol anesthesia: A systematic review with meta-analysis and trial sequential analysis. Journal of Advanced Research. 2020;**24**:223-238. DOI: 10.1016/j.jare.2020.02.008

[63] Hirooka K, Ukegawa K, Nitta E, Ueda N, Hayashida Y, Hirama H, et al. The effect of steep Trendelenburg positioning on retinal structure and function during robotic-assisted laparoscopic procedures. Journal of Ophthalmology. 2018;**2018**:1027397. DOI: 10.1155/2018/1027397

[64] Hirooka K, Ukegawa K, Nitta E, Ueda N, Hayashida Y, Hirama H, et al. Retinal nerve fiber layer thickness progression after robotic-assisted laparoscopic radical prostatectomy in glaucoma patients. Journal of Ophthalmology. 2019;**2019**:6576140. DOI: 10.1155/2019/6576140

[65] Awad H, Bai M, Ramadan ME, Shabsigh A, Backes F, Craven MA, et al. The effect of increased intraocular pressure during steep Trendelenburg positioning in robotic prostatectomy and hysterectomy on structural and functional ocular parameters. Anesthesia and Analgesia. 2020;**130**(4):975-982. DOI: 10.1213/ANE.0000000000004547

[66] Balkan B, Emir NS, Demirayak B, Çetingok H, Bayrak B. The effect of robotic surgery on intraocular pressure and optic nerve sheath diameter: A prospective study. Brazilian Journal of

Anesthesiology. 2021;**71**(6):607-611. DOI: 10.1016/j.bjane.2021.02.035

[67] Kondo Y, Echigo N, Mihara T, Koyama Y, Takahashi K, Okamura K, et al. Intraocular pressure during robotic-assisted laparoscopic prostatectomy: A prospective observational study. Brazilian Journal of Anesthesiology. 2021;**71**(6):618-622. DOI: 10.1016/j.bjane.2021.02.041

[68] Yoo YC, Shin S, Choi EK, Kim CY, Choi YD, Bai SJ. Increase in intraocular pressure is less with propofol than with sevoflurane during laparoscopic surgery in the steep Trendelenburg position. Canadian Journal of Anaesthesia. 2014; **61**(4):322-329. DOI: 10.1007/s12630-014-0112-2

[69] Calza E, Porpiglia F, Fiori C, Giannone A, Pusineri A, Meli A. Intraocular pressure increase during robot assisted laparoscopic prostatectomy: Balanced anesthesia versus total intravenous anesthesia. In: Annual Meeting of the American Urological Association, AUA. Washington, DC, United States: Elsevier; 2011, Conference 14 to 19 May 2011

[70] Matsuyama K, Fujinaka W, Takatori M. Intra-occular pressure during robotic-assisted laparoscopic radical prostectomy. Masui. 2014;**63**(12): 1366-1368. Japanese

[71] Hwang JW, Oh AY, Hwang DW, Jeon YT, Kim YB, Park SH. Does intraocular pressure increase during laparoscopic surgeries? It depends on anesthetic drugs and the surgical position. Surgical Laparoscopy, Endoscopy & Percutaneous Techniques. 2013;**23**(2):229-232. DOI: 10.1097/SLE.0b013e31828a0bba

[72] Kaur G, Sharma M, Kalra P, Purohit S, Chauhan K. Intraocular pressure changes during laparoscopic surgery in Trendelenburg position in patients anesthetized with Propofol-based total intravenous anesthesia compared to sevoflurane anesthesia: A comparative study. Anesthesia, Essays and Researches. 2018;**12**(1):67-72. DOI: 10.4103/aer.AER_177_17

[73] Seo KH, Kim YS, Joo J, Choi JW, Jeong HS, Chung SW. Variation in intraocular pressure caused by repetitive positional changes during laparoscopic colorectal surgery: A prospective, randomized, controlled study comparing propofol and desflurane anesthesia. Journal of Clinical Monitoring and Computing. 2018;**32**(6):1101-1109. DOI: 10.1007/s10877-018-0116-5

[74] Balkan B, Tontu F, Genç Moralar D, Demirayak B, Emir NS, Yektaş A. The effect of anesthetic agents on intraocular pressure during laparoscopic gynecological surgery performed in the Trendelenburg position: A randomized clinical trial. Ulusal Travma ve Acil Cerrahi Dergisi. 2022;**28**(4):498-507. DOI: 10.14744/tjtes.2020.56019

[75] Chang CY, Chen HA, Chien YJ, Wu MY. Attenuation of the increase in intraocular pressure with dexmedetomidine: Systematic review with meta-analysis and trial sequential analysis. Journal of Clinical Anesthesia. 2021;**68**: 110065. DOI: 10.1016/j.jclinane.2020.110065

[76] Kim NY, Yoo YC, Park H, Choi YD, Kim CY, Bai SJ. The effect of dexmedetomidine on intraocular pressure increase in patients during robot-assisted laparoscopic radical prostatectomy in the steep Trendelenburg position. Journal of Endourology. 2015;**29**(3):310-316. DOI: 10.1089/end.2014.0381

[77] Joo J, Koh H, Lee K, Lee J. Effects of systemic administration of

dexmedetomidine on intraocular pressure and ocular perfusion pressure during laparoscopic surgery in a steep Trendelenburg position: Prospective, randomized, double-blinded study. Journal of Korean Medical Science. 2016; **31**(6):989-996. DOI: 10.3346/jkms.2016.31.6.989

[78] Kitamura S, Takechi K, Nishihara T, Konishi A, Takasaki Y, Yorozuya T. Effect of dexmedetomidine on intraocular pressure in patients undergoing robot-assisted laparoscopic radical prostatectomy under total intravenous anesthesia: A randomized, double blinded placebo controlled clinical trial. Journal of Clinical Anesthesia. 2018;**49**:30-35. DOI: 10.1016/j.jclinane.2018.06.006

[79] Kumari P, Deepshikha KA, Kumar R, Kumar A, Surabhi. Fixed dilated pupils following laparoscopic urological procedure: Steep Trendelenburg position was the cause. Saudi Journal of Anaesthesia. 2021;**15**(4):469-470. DOI: 10.4103/sja.sja_403_21

[80] Raz O, Boesel TW, Arianayagam M, Lau H, Vass J, Huynh CC, et al. The effect of the modified Z Trendelenburg position on intraocular pressure during robotic assisted laparoscopic radical prostatectomy: A randomized, controlled study. The Journal of Urology. 2015;**193**(4):1213121-1213129. DOI: 10.1016/j.juro.2014.10.094

[81] Nishikawa M, Watanabe H, Kurahashi T. Effects of 25- and 30-degree Trendelenburg positions on intraocular pressure changes during robot-assisted radical prostatectomy. Prostate International. 2017;**5**(4):135-138. DOI: 10.1016/j.prnil.2017.03.008

[82] Molloy B. A preventive intervention for rising intraocular pressure: Development of the Molloy/Bridgeport anesthesia associates observation scale. AANA Journal. 2012;**80**(3):213-222

[83] You AH, Song Y, Kim DH, Suh J, Baek JW, Han DW. Effects of positive end-expiratory pressure on intraocular pressure and optic nerve sheath diameter in robot-assisted laparoscopic radical prostatectomy: A randomized, clinical trial. Medicine (Baltimore). 2019;**98**(14): e15051. DOI: 10.1097/MD.0000000000015051

[84] Tosh P, Krishnankutty SV, Rajan S, Nair HM, Puthanveettil N, Kumar L. Does restrictive fluid strategy during robotic pelvic surgeries obtund intraoperative rise in intraocular pressure? Anesthesia, Essays and Researches. 2018;**12**(1):155-158. DOI: 10.4103/aer.AER_144_17

[85] Alexander I, Ahmad S, Inbar BS, Joseph P. The effect of perioperative fluid management on intraocular pressure during gynecologic laparoscopic pelvic surgery. Journal of Ophthalmology. 2018;**2018**:1457851. DOI: 10.1155/2018/1457851

[86] Joo J, Kim J, Lee J. Effect of continuous systemic administration of esmolol on intraocular pressure during surgery in a sustained steep Trendelenburg position. Journal of Glaucoma. 2017;**26**(12):1068-1071. DOI: 10.1097/IJG.0000000000000746

[87] Mathew DJ, Greene RA, Mahsood YJ, Hallaji N, Vargas AMB, Jin YP, et al. Preoperative brimonidine tartrate 0.2% does not prevent an intraocular pressure rise during prostatectomy in steep Trendelenburg position. Journal of Glaucoma. 2018; **27**(11):965-970. DOI: 10.1097/IJG.0000000000001047

[88] Roth S, Moss HE, Vajaranant TS, Sweitzer B. Perioperative care of the

patient with eye pathologies undergoing nonocular surgery. Anesthesiology. 2022;**137**(5):620-643. DOI: 10.1097/ALN.0000000000004338

[89] Lee M, Dallas R, Daniel C, Cotter F. Intraoperative management of Increased intraocular pressure in a patient with glaucoma undergoing robotic prostatectomy in the Trendelenburg position. A A Case Report. 2016;**6**(2): 19-21. DOI: 10.1213/XAA.0000000000000213

[90] Moriyama Y, Miwa K, Yamada T, Sawaki A, Nishino Y, Kitagawa Y. Intraocular pressure change during laparoscopic sacral colpopexy in patients with normal tension glaucoma. International Urogynecology Journal. 2019;**30**(11):1933-1938. DOI: 10.1007/s00192-018-03866-w

[91] Gayer S, Gedde SJ. Intraoperative management of increased intraocular pressure in a patient with glaucoma undergoing robotic prostatectomy in Trendelenburg position. A A Case Report. 2016;**6**(2):17-18. DOI: 10.1213/XAA.0000000000000233

[92] Nuzzi R, Tridico F. Ocular complications in laparoscopic surgery: Review of existing literature and possible prevention and treatment. Seminars in Ophthalmology. 2016;**31**(6):584-592. DOI: 10.3109/08820538.2015.1009557

[93] Lee LA, Stoelting RK. APSF-sponsored conference on perioperative visual loss develops consensus conclusions. APSF Newsletter. 2013;**27**: 52-53

[94] Aceto P, Beretta L, Cariello C, Claroni C, Esposito C, Forastiere EM, et al. Joint consensus on anesthesia in urologic and gynecologic robotic surgery: Specific issues in management from a task force of the SIAARTI, SIGO, and SIU. Minerva Anestesiologica. 2019; **85**(8):871-885. DOI: 10.23736/S0375-9393.19.13360-3

[95] Roth S. Perioperative visual loss: What do we know, what can we do? British Journal of Anaesthesia. 2009;**103** (Suppl. 1):i31-i40. DOI: 10.1093/bja/aep295

[96] Postoperative Visual Loss Study Group. Risk factors associated with ischemic optic neuropathy after spinal fusion surgery. Anesthesiology. 2012; **116**(1):15-24. DOI: 10.1097/ALN.0b013e31823d012a

Chapter 4

Intraocular Pressure Measurement in Africa: A Review of Literature

Thokozani Mzumara and Owen Banda

Abstract

Ocular hypertension (OHT) is a pervasive condition easily mistaken for glaucoma. In Africa, glaucoma is reported the highest, hence there is a need to properly distinguish it from ocular hypertension, which is the rise in IOP in the absence of glaucomatous changes. Many methods exist for measuring IOP; however, recent focus has been on non-invasive techniques. This review aims to assess the measurement of IOP among non-glaucomatous Africans. The research adopts a systematic approach employing the five-step framework by Arksey and O Malley. We used the research question to formulate a search strategy used to extract the studies included in the review. Next, we used keywords in combination with Boolean operators to search the PubMed database. The study analyzed articles published in English between 2010 and 2023. The search identified 136 articles. Both authors checked the article for screening and eligibility. The instruments used include GAT, perking's tonometry, rebound tonometry, tonopachy, and the value of IOP among Africans ranges from 11 to 16 mmHg and varies according to instruments and population. There is a wide variation in the value of IOP among Africans. Practitioners should consider the population mean for each instrument used and central corneal thickness during IOP measurements.

Keywords: intraocular pressure, Africa, ocular hypertension, tonometry, nonglaucoma, ocular biometry, Africa

1. Introduction

Glaucoma is a major cause of irreversible blindness and a major public health concern on the African continent [1]. Intraocular pressure (IOP) is the only modifiable factor of glaucoma and ocular hypertension, which is regarded among other risk factors such as thinner corneas, older age, and greater cup-to-disk ratio [2]. Ocular hypertension (OHT) refers to eyes in which the IOP is higher than the normal range for the general population but is free of glaucomatous damage [3]. Noteworthy, the majority of people with OHT do not progress to glaucoma [2], nevertheless proper distinction is required to clinically detect OHT and glaucoma since glaucoma causes mechanical damage to the optical nerve thereby disrupting visual signals sent to the brain thereby affecting vision. OHT represents 75% of all forms of glaucoma [3].

The prevalence of OHT differs across populations. For instance, the cut-off value for OHT is 20.4 mmHg, 21.5 mmHg, and 22.6 mmHg for the Chinese, Indian, and Malay population groups [4]. However, OHT is highest among Africans compared to other continents, reflecting the effect of race and heredity [5]. Noteworthy, IOP measurements vary based on measurement technique and population. In general, the upper limit for "normal" IOP is two standard deviations above the population mean. Needless to say, practitioners need to consider the normal values during diagnosis and follow-up of glaucoma suspects and patients concerning measurement technique and population [6].

Despite that black people have thinner corneas, which predispose them to the risk of developing OHT and glaucoma [2], information regarding the normative values of different methods of IOP among people of African descent remains unknown. In addition, there are diverse ethnic groups in Africa, which calls for a need to assess the geographical distribution of IOP across the region. This chapter will summarize the current data on normative values of IOP including the current techniques employed for measuring IOP in Africa focusing on the challenges and opportunities associated with various measurement techniques employed in the region.

The paper is split into three sections. The first section will address the methodologies employed to answer the research question in this review, including the research design, search strategy, data retrieval, and analytical strategies. Next, the paper presents the results of the search summarizing the data retrieved in the previous section. Finally, the paper discusses the research findings in light of the available literature considering the situation of the African eye health care landscape discussing the advantages and disadvantages of using the diagnostic technique considering this setting.

2. Methods

The review adopted a systematic review approach postulated by Arksey and O'Malley [7]. The steps include formulating the research questions, identifying studies, selecting relevant studies, charting the data, and collating, summarizing, and reporting the findings.

2.1 Identifying research questions

The review is guided by the following research question.

- What is the distribution of intraocular pressure in nonglaucomatous Africans?

2.2 Eligibility of research studies

To assess the eligibility of the research question formulated, we employed the Population Concept Context framework as illustrated below

- Population: nonglaucomatous patients
- Concept: intraocular pressure or ocular hypertension
- Context: Africa

2.3 Identifying relevant studies

2.3.1 Inclusion criteria

As an inclusion criterion, we selected studies published from 2010 to 2023. In addition, we retrieved studies from the citations of our searched articles. Only studies published in the English language and conducted among nonglaucomatous Africans were entered into the review.

2.3.2 Exclusion criteria

We excluded studies conducted among nonhuman subjects. Furthermore, studies conducted to develop new instruments were not included in the review. In addition, we removed studies employing the following research designs retrospective and systematic reviews.

2.3.3 Electronic database search

We searched PubMed and Google Scholar databases for research on ocular hypertension or intraocular pressure among African populations. We used the following keywords and a combination of Boolean operators' ocular hypertension OR intraocular pressure AND Africa AND nonglaucoma.

A manual search through the references of retrieved studies was conducted and all studies that satisfied the inclusion criteria were included in the review.

2.3.4 Selecting relevant studies

The remaining studies were assessed by reviewing the title and abstract. This was done by members of the study and they reached a consensus. When they failed to reach a consensus, a third independent reviewer was invited to help decide. The selected papers underwent review using a critical appraisal tool for cross-sectional studies to ensure quality studies are included in the review (**Figure 1**).

2.3.5 Charting the data

We included relevant information, including the author's name, year of publication, county of study, methods and design of the study, mean IOP, and standard deviation. The process was conducted by the two authors and meetings were held to compare the findings and agree on the charting.

2.3.6 Collating, summarizing, and reporting the findings

The charted data was summarized to communicate the values of IOP. The report followed the guidelines included in the Preferred Reporting Items for Systematic Reviews and Metanalysis (PRISMA) [7].

3. Results

The selected studies are illustrated in **Table 1**. Out of the 54 studies identified, 46 were excluded because they did not meet the inclusion criteria. About eight

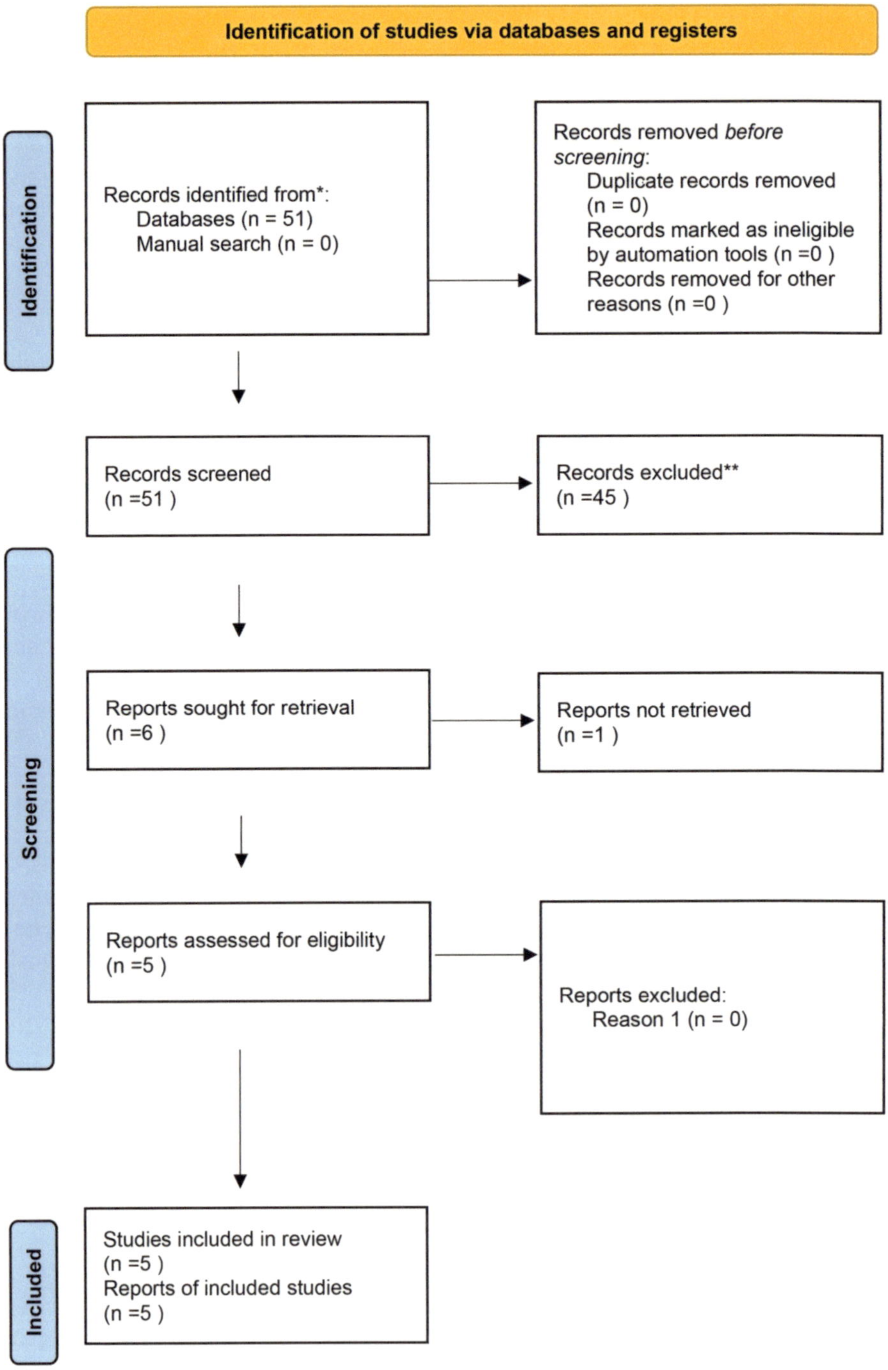

Figure 1.
Prisma.

were sought for retrieval and only five out of eight were screened for eligibility and included in the qualitative synthesis.

Out of the five studies included in the synthesis, two were from Nigeria [9, 10], 1 from Ethiopia [8], 1 from South Africa [12], and 1 from Egypt [11].

Author (year)	Country	Methods	Mean IOP (standard deviation)	Technique usage
[8]	Ethiopia	Prospective cross-sectional study	12.1 mmHg (SD = 5.0 mmHg)	Rebound tonometer (iCare)
[9]	Southwest Nigeria	Comparative cross-sectional study	• 15.18 ± 4.26 mmHg (GAT) • 16.32 ± 4.48 mmHg (Rebound iCare tonometer)	GAT & iCare
[10]	Southwest Nigeria	Hospital-based case control study	14.2 ± 2.6 mmHg	GAT
[11]	Egypt	Prospective cross-sectional study	11.5 ± 2.34 mmHg	Perkins tonometer
[12]	South Africa	Cross-sectional study	• 14.79 ± 3.09 mmHg (GAT) • 14.32 ± 3.57 mmHg (Tonopachy) • 16.64 ± 4.38 mmHg (iCare)	GAT, Tono-pachymeter (Tonopachy) & iCare

Table 1.
An overview of reviewed studies.

3.1 IOP variation according to instruments

The reviewed studies showed that there are differences in the mean IOP based on the instruments used. For instance, Ashano et al. [9] found that IOP measured with a rebound tonometer differed significantly from the mean IOP measured with GAT although the two were significantly correlated. The I Care rebound tonometer shows higher results compared with GAT measured by one practitioner on the same set of patients. In agreement, the South African study [12] found that tonopachy and GAT findings were in tandem while the rebound tonometer produced higher IOP2021 values. Therefore, the two procedures cannot be used interchangeably. Nevertheless, the rebound tonometer is relatively easy to use as it requires minimum training, consumes less time and does not need anapestic eye drops, and is portable ideal for outreach programs [8].

3.2 Central corneal thickness and IOP

Furthermore, it was discovered that there is a positive correlation between central corneal thickness and intraocular pressure as evidenced by an increase in IOP associ-ated with CCT in a study in Nigeria [10]. In contrast, a study conducted in South Africa [12] found no relationship between CCT and IOP measured with GAT, rebound tonometer and tonopachy. This relationship highlights the need to exercise caution with thinner corneas which may underestimate IOP and delay initiation of treatment among the continental Africans.

4. Discussion

The chapter summarizes the common methods used to measure the modifiable risk factor of OHT and glaucoma namely IOP in the African setting. The chapter has achieved this by systematically reviewing research that has been conducted on the African continent to assess IOP among normal populations. The following narrative provides some of the common themes emerging from the literature on the topic.

4.1 Instruments used to measure IOP

This review has discovered that the most common instruments used for measuring IOP are Goldman applanation tonometry, Perkins tonometer, and rebound tonometry. The results of this paper are not surprising considering that GAT is considered the gold standard of measurement worldwide. On the other hand, Perkins and eye care could be mostly utilized for their compatibility. Unlike the Goldman applanation tonometry, which requires sophisticated equipment such as slit lamp biomicroscope, the Perkins and rebound tonometry are portable and therefore easy to use during outreach programs. In most African regions, access to eye care services is complimented by community outreach programs, especially through cataract camps. The major drawback of GAT also lies in the requirement of fluorescein and local anesthesia as part of diagnostic drugs. This becomes a challenge in the African context against the backdrop of insufficient and unavailability of drugs and medicines coupled with issues of pilferage. In Africa, the delivery of eye care services continues to be undermined by health system performance bottlenecks [13]. The medical practice in Africa is constrained by the unavailability of equipment, a lack of diagnostic skills, and expensive treatment regimens [1].

4.2 IOP and central corneal thickness

The review found that there is a discrepancy in the relationship between IOP and central corneal thickness. This also concurs with Osman et al. [14] who postulated that corneal thickness and irregularities may affect GAT. We attribute the differences to different corneal thickness parameters across populations. Accordingly, ocular parameters differ significantly with ethnicity. The findings of this review suggest that practitioners in Africa should consider the corneal thickness when measuring IOP. IOP measurements are heavily dependent on corneal parameters and can be a major source of error. However, accuracy in IOP measurements is integral [15].

The range of CCT across Africa is 519–550 μm. The highest CCT value was reported among a Nigerian sample (550 μm) [16]. The lowest was reported among the Ethiopian and South African populations (519 μm) [12, 17]. Instruments such as the Keeler Pulsair Eye non-contact tonometer are prone to changes in corneal thickness and should be used together with CCT [18]. Modern techniques allow for the CCT measurements to be incorporated in the calculation of the final result of IOP measurement. For example, the non-contact automated Reichert's Ocular Response Analyzer (ORA; Reichert, Inc., Depew, NY) produces two measures of IOP: the Goldman correlated IOP and the corneal compensated IOP. Moreover, it assesses corneal resistance factor and corneal hysteresis, two biomechanical properties [19].

5. Summary

In conclusion, the chapter has reviewed the current techniques of measuring intraocular pressure that have been applied on the African continent. There is a paucity of population-based studies, especially with a nationally representative sample measuring the distribution of IOP in Africa. The majority of studies conducted are comparative studies. The commonly used techniques include GAT and handheld devices such as Perkins tonometry and rebound tonometry. Despite the scarcity of these instruments mainly associated with high purchasing prices, they offer reliable methods of monitoring ocular hypertension and glaucoma. African eye care practitioners should familiarize themselves with the methods and their drawbacks while exploring novel techniques to aid in the fight against ocular hypertension and glaucoma.

Author details

Thokozani Mzumara[1,2*] and Owen Banda[1]

1 Department of Optometry, Mzuzu University, Mzuzu, Malawi

2 Department of Ophthalmology, Mzuzu Urban Health Centre, Mzimba North District Hospital, Mzuzu, Malawi

*Address all correspondence to: mzumarathokozani@gmail.com

References

[1] Kyari F et al. The current status of glaucoma and glaucoma care in sub-Saharan Africa. The Asia-Pacific Journal of Ophthalmology. 2018;7(6):375-386

[2] Friedman DS, Wilson M, Liebmann JM, Fechtner RD, Weinreb RN. An evidence-based assessment of risk factors for the progression of ocular hypertension and glaucoma. American Journal of Ophthalmology. 2004;**138**(3):19-31. DOI: 10.1016/j.ajo.2004.04.058

[3] Gordon MO, Kass MA, for the Ocular Hypertension Treatment Study Group. The ocular hypertension treatment study: Design and baseline description of the participants. Archives of Ophthalmology. 1999;**117**(5):573-583. DOI: 10.1001/archopht.117.5.573

[4] Thakur S, Lavanya R, Yu M, Tham YC, Da Soh Z, Teo ZL, et al. Six-year incidence and risk factors for primary open-angle glaucoma and ocular hypertension: The Singapore epidemiology of eye diseases study. Ophthalmology Glaucoma. 12 Aug 2023:S2589-4196(23)00157-6. DOI: 10.1016/j.ogla.2023.08.003. PMID: 37574187. Epub ahead of print

[5] Sharif NA. Therapeutic drugs and devices for tackling ocular hypertension and glaucoma, and need for neuroprotection and cytoprotective therapies. Frontiers in Pharmacology. 2021;**12**:729249. DOI: 10.3389/fphar.2021.729249

[6] Georgopoulos G, Andreanos D, Liokis N, Papakonstantinou D, Vergados J, Theodossiadis G. Risk factors in ocular hypertension. European Journal of Ophthalmology. 1997;7(4):357-363. DOI: 10.1177/112067219700700409

[7] Abraham CH, van Staden D, Rampersad N. Barriers and enablers to low vision care and rehabilitation in sub-Saharan Africa within a global context. Clinical and Experimental Optometry. 2023:1-11

[8] McClain IJ, Rooney DM, Tabin GC. Intraocular pressure screening during high-volume cataract surgery outreach in Ethiopia. BMC Ophthalmology. 2022;**22**(1):397. DOI: 10.1186/s12886-022-02618-1

[9] Ashano O, Oderinlo O, Ogunro A, Ashano E. A comparison of goldmann applanation tonometry and rebound tonometry measurements among patients attending a glaucoma clinic in Southwest Nigeria. Nigerian Journal of Clinical Practice. 2022;**25**(9):1542-1547. DOI: 10.4103/njcp.njcp_95_22

[10] Oluwaniyi AT, Olawoye O, Sarimiye TF, Ajayi BGK. Comparison of ocular biomery in primary open angle glaucoma and non-glaucoma in south West Nigeria. Journal of the West African College of Surgeons. 2023;**13**(2):37-44. DOI: 10.4103/jwas.jwas_264_22v

[11] Moussa IR, Kassem RR, Edris NA, Khalil DH. Normal intraocular pressure in Egyptian children and meta-analysis. Eye. 2022;**36**(6):1266-1273. DOI: 10.1038/s41433-021-01633-8

[12] Rampersad N, Mashige KP, Jhetam S. A comparison of intraocular pressure values obtained with the Tono-pachymeter NT530P, iCare® rebound tonometer and Goldmann applanation tonometer. South African Optometrist. 2011;**70**:109-116

[13] Bechange S, Jolley E, Virendrakumar B, Pente V, Milgate J,

Schmidt E. Strengths and weaknesses of eye care services in sub-Saharan Africa: A metasynthesis of eye health system assessments. Health Services Research. 2020;**20**:381. DOI: 10.1186/s12913-020-05279-2

[14] Osman EA, Gikandi PW, Al-Jasser AA, Alotaibi M, Mousa A. Comparison of Goldmann applanation, noncontact air puff, and Tono-pen XL tonometry in normal controls versus glaucoma patients at a University Hospital in Riyadh, Saudi Arabia. Middle East African Journal of Ophthalmology. 2018;**25**(1):8-13. DOI: 10.4103/meajo.MEAJO_291_16

[15] Davey PG, Elsheikh A, F, D. Clinical evaluation of multiparameter correction equations for Goldmann applanation tonometry. Eye. 2013;**27**(5):621-629. DOI: 10.1038/eye.2013.23

[16] Iyamu E, Eze NM. The relationship between central corneal thickness and corneal curvature in adult Nigerians. South African Optometrist. 2011;**70**:44-50

[17] Gelaw Y, Kollmann M, Irunga NM, Ilako DR. The influence of central corneal thickness on intraocular pressure measured by Goldmann applanation tonometry among selected Ethiopian communities. Journal of Glaucoma. 2010;**19**:514-518. DOI: 10.1097/IJG.0b013e3181ca7708

[18] Babalola OE, Kehinde AV, Iloegbunam AC, Akinbinu T, Moghalu C, Onuoha I. A comparison of the Goldmann applanation and non-contact (Keeler Pulsair EasyEye) tonometers and the effect of central corneal thickness in indigenous African eyes. Ophthalmic and Physiological Optics. 2009;**29**(2):182-188. DOI: 10.1111/j.1475-1313.2008.00621.x

[19] McCann P, Hogg RE, Wright DM, McGuinness B, Young IS, Kee F. Comparison of Goldmann applanation and ocular response analyser tonometry: Intraocular pressure agreement and patient preference. Eye. 2020;**34**(3):584-590. DOI: 10.1038/s41433-019-0556-2

Chapter 5

Pharmacologic and Natural Therapeutics in Glaucoma Management

Karen Allison, Kevin Morabito Jr, Deepkumar Patel and Brandon W. Montoya

Abstract

Glaucoma is the leading cause of irreversible blindness worldwide. As diseased population continues to grow, it is important to review both the well-established and recently developed therapeutics available today to best treat this ocular condition. This chapter will discuss the pharmacologic therapies most commonly used to lower intraocular pressure (IOP) in primary open angle glaucoma patients. It will also examine both natural agents and lifestyle modifications that have been shown to have an effect on intraocular pressure. The prostaglandin analog latanoprost, continues to be the most widely accepted first line medication used to treat glaucoma. However, the efficacious, recently developed, Rho-kinase inhibitor Netarsudil, and fixed dose combination of Netarsudil-Latanoprost should continue to increase in utilization. Multiple mechanisms are often used together to treat glaucoma. Fixed dose combination drug therapy has the potential to decrease patient burden, increase compliance, and improve clinical outcomes.

Keywords: glaucoma, intraocular pressure, pharmacologic, natural, therapeutics

1. Introduction

Glaucoma is a term used to describe a group of diseases characterized by damage to the optic nerve and retinal nerve fiber layer. It is a chronic progressive optic neuropathy that causes peripheral and occasionally central vision loss. It is the leading cause of irreversible blindness worldwide [1]. The number of people affected by glaucoma is anticipated to continue to rise. It is estimated that approximately 3 million people in the US currently have glaucoma, and this number is expected to grow to 6.3 million by 2050 [2]. Globally, it is estimated that 76 million people suffer from glaucoma with a projected growth to 112 million by 2040 [2]. With such stark projections, it is clear that developments in therapeutics will need to progress in order to optimize patient care and clinical outcomes. Additionally, glaucoma disproportionately affects Black populations in the US. It is the most common cause of blindness in Black persons with a prevalence of 6.1% [2]. The prevalence in Latino communities is second highest at 4.1%, followed by Asian Americans at 3.5% and non-Hispanic White persons at 2.8% [2]. These

 IntechOpen

statistics point to a need to increase the screening and access to care for persons from historically medically underserved communities.

There is no cure for glaucoma. Intraocular pressure (IOP) is the only known modifiable risk factor and therefore of the utmost importance in controlling disease progression. However, in many patients intraocular pressure is only slightly elevated or still within the normal range [1].

The rise in pressure associated with glaucoma can be painless and is often unnoticeable to the patient until subsequent symptoms are present. Early detection is extremely important because any visual field loss noticed by the patient is irreversible. Other general risk factors include older age, race and ethnicity, and family history of glaucoma.

All pharmacologic therapeutics used to treat glaucoma work by lowering intraocular pressure. The common goal is to decrease the loss of ganglion cells, thinning of the retinal nerve fiber layer, and cupping of the optic disc to slow progression. There are a variety of medications used today that work through several different mechanisms. These include cholinergic agents, alpha adrenergic agonists, beta blockers, carbonic anhydrase inhibitors, nitric oxides, prostaglandin analogs, and rho-kinase inhibitors. Additionally, there is a very limited amount of data from studies focusing on natural treatment modalities, therefore the emergence of a potential natural therapy or substance that has the ability to reduce IOP in order to slow glaucomatous progression would be extremely valuable.

2. Treatment

Despite current knowledge of risk factors, genetics, and pathophysiology, the only medical form of treatment for primary open angle glaucoma (POAG) today remains to be intraocular pressure lowering medications. In particular cases, surgical procedures are recommended. However, as the focus of this chapter is pharmacologic and naturopathic therapy, they will only be covered briefly. At this time, there is no way to cure POAG or OHT. The goals of treatment are to delay the progression of visual field loss for the duration of the patient's lifetime.

IOP lowering medications have been established as the cornerstone of glaucoma treatment for decades. The Ocular Hypertension Treatment Study established that topical ocular hypotensive medication is effective in delaying or preventing the onset of POAG in individuals with elevated IOP [3]. In the study, participants with no evidence of glaucomatous damage were randomized into 2 groups. They were placed in either a treatment with a topical ocular hypertensive medication group or an observation group. The goals of the treatment were a 20% reduction in intraocular pressure and an intraocular pressure of less than or equal to 24 mm Hg. During the course of the study, the mean ± standard deviation of the reduction of IOP in the medication group was 22.5 ± 9.9%. In the observation group, IOP declined by 4.4 ± 11.6% [3]. Additionally, at 60 months, the cumulative probability of developing POAG was 4.4% in the medication group and 9.5% in the observation group. The protective effect of treatment was statistically significant for both optic disc and visual field changes [3]. Furthermore, this study, along with the Early Manifest Glaucoma Trial and the United Kingdom Glaucoma Treatment Study indicated that the degree of reduction of pressure influences disease progression [4, 5]. The Early Manifest Glaucoma Trial estimated that each 1 mm Hg reduction in intraocular pressure reduced the risk of glaucoma progression by about 10% [6]. It can be concluded that reducing IOP through topical ocular hypotensive medications is an extremely effective treatment for delaying disease progression and improving clinical outcomes.

There are numerous forms of medication used to reduce intraocular pressure in POAG patients. These include nitric oxides, cholinergic agents, alpha adrenergic agonists, beta blockers, carbonic anhydrase inhibitors, prostaglandin analogs, and rho kinase inhibitors. It is important to address that personalized treatment promotes better patient outcomes in POAG management. A common strategy is to recommend lowering IOP to a specific target pressure. At this particular pressure, the rate of disease progression should be slowed enough to prevent further damage. Generally, target IOP aims to be a reduction of 20–50% from baseline IOP [7]. This number is established from factors including pre-treatment baseline IOP, severity of visual field loss, risk of progression, life expectancy, and potential for adverse effects [7].

Laser or incisional surgeries are indicated when medical treatment is unable to reduce IOP to target pressure or a patient is unable to tolerate pharmacotherapy. In severe cases and when a patient does not adhere to medical recommendations, surgery can be preferred as a first line therapy [8]. Laser trabeculoplasty is a common form of treatment in which a laser induces biological changes in the trabecular meshwork. This is able to increase aqueous outflow and reduce IOP. Trabeculectomy is the most commonly performed incisional surgery used to lower intraocular pressure [8]. It consists of the excision of a small portion of the trabecular meshwork and or adjacent corneoscleral tissue. This provides a drainage route for aqueous humor from within the eye to underneath the conjunctiva where it is absorbed [8]. Additionally, MIGS procedures, or microinvasive glaucoma surgeries are a relatively new treatment option. Most MIGS work by dilating, cleaving open, or bypassing tissue that is obstructing aqueous outflow, or by inserting a device into an outflow structure to increase drainage [2]. Each procedure comes with its own set of risks. Often, a combination of surgery and pharmacologic therapeutic treatments can improve patient outcomes. In this chapter we will be focusing on pharmacologic and non-pharmacologic therapies.

3. Cholinergic agents

Cholinergic agents, also known as parasympathomimetics, are a useful tool in treating primary open angle glaucoma. The miotic effects of cholinergic agents were first reported by Thomas Fraser in 1862 after studying the active ingredient of a potentially lethal poison used in West Africa derived from the Calabar bean – physostigmine [9]. Pilocarpine was discovered to be less toxic and became the first known pharmacotherapy used to treat glaucoma in 1877 [9]. Pilocarpine and carbachol are two prominent examples, however pilocarpine is the best known of this class of IOP lowering drugs. Pilocarpine works by inducing smooth muscle contraction of the cells in the ciliary body. This leads to an increase in aqueous humor outflow through the trabecular meshwork pathway by widening the trabecular meshwork and Schlemm's canal [10]. This can lead to an IOP reduction of 20–25% [10]. Unfortunately, cholinergic agents are known to produce a wide range of systemic side effects and are not well tolerated by patients. They are even contraindicated in patients under 40 years of age [10]. This is likely due to overstimulation of muscarinic acetylcholine receptors, inducing bradycardia, negative cardiac inotropy, salivation, sweating, and gastrointestinal stimulation [9]. Ocular adverse effects include retinal detachment, ciliary cramps, increased pupillary block, and blurred vision due to induced myopia [7, 10]. This array of side effects typically restricts the use of pilocarpine to short term and very limited situations.

4. Alpha adrenergic agonists

Brimonidine, which is sold under the brand name Alphagan, and apraclonidine, are the most common alpha adrenergic agonists used to treat primary open angle glaucoma. Alpha adrenergic agonists operate by stimulating alpha receptors to release norepinephrine [10]. Norepinephrine is the main neurotransmitter of the adrenergic system. As an alpha-2 agonist, brimonidine is useful in causing vasoconstriction in the ciliary body. This will result in a decrease in aqueous humor production and consequential lowering of intraocular pressure [10]. However, after chronic use, the mechanism of action shifts to the uveoscleral pathway. This mechanism contracts the ciliary body and increases outflow [11]. Brimonidine can cause a 20–25% reduction in intraocular pressure, making it a useful IOP lowering agent [7]. Although generally more well received than other alpha adrenergic receptor agonists, brimonidine is associated with several adverse effects. These include contact dermatitis, ocular irritation, allergic conjunctivitis, anterior uveitis, hyperemia, fatigue, dizziness, and hypotension [7, 10]. Of particular importance in glaucoma treatment is the potential for neuroprotective properties. As there are alpha-2 receptors present in retinal ganglion cells, Brimonidine also has the potential to reduce retinal ganglion cell loss in glaucoma. There have been several studies involving rat models to examine this hypothesis, with one concluding that Brimonidine was in fact useful in significantly reducing retinal ganglion cell loss [10]. With the potential for significant IOP reduction and neuroprotection, but with occasional adverse effects, Brimonidine is often considered a second line glaucoma treatment. It is also frequently used in combination with other topical therapies.

5. Beta adrenergic antagonists

Beta receptor antagonists, or beta blockers as they are commonly known, are a significant and effective form of treatment for primary open angle glaucoma. Beta receptors are located throughout the eye and their antagonists operate through a mechanism of reducing aqueous humor production in the ciliary body. This is achieved through decreasing cAMP production [10]. Like other effective IOP lowering medications, beta blockers are able to reduce intraocular pressure 20–25% of their initial values [7]. Timolol, brand name Timoptic and Betaxolol, brand name Kerlone, are some of the more frequently prescribed options. Timolol was the first nonselective beta blocker used to treat glaucoma on the market and was the most used anti glaucoma drug for many years [10]. As with all pharmacologic therapies, there are potential adverse effects associated with beta blockers. Some severe pulmonary and cardiovascular systemic side effects can present with the use of non-specific beta receptor antagonists. This contraindicates their prescription for patients who suffer from COPD, asthma, decompensated chronic heart failure, symptomatic bradycardia or heart block, and a history of syncope without diagnosis [10]. There is also concern about the systemic hypotensive effect of beta blockers. They have the potential to reduce blood flow to the optic nerve, which could result in exacerbating glaucomatous progression [9]. Betaxolol was developed with the hope of less adverse effects and as a selective beta-1 adrenergic blocker, and it may have less of a tendency to induce bronchospasm [9]. In addition, beta blockers may also have some neuroprotective effects. One study concluded that Timolol was able to reduce IOP and protect retinal ganglion cells in an experimental rat model through upregulation of brain derived neurotrophic factor [12]. The dangerous side effects that can occur with beta blocker use require them to be prescribed with caution. However, their IOP lowering effects and potential for neuroprotection make them a valuable instrument in the treatment of primary open angle glaucoma, either alone or in combination with other agents.

6. Carbonic anhydrase inhibitors

Carbonic anhydrase inhibitors are a staple of primary open angle glaucoma treatment. Carbonic anhydrase is important for the production of aqueous humor and through its inhibition, a reduction in aqueous humor can cause a decrease in intraocular pressure. Through restricting the amount of sodium and bicarbonate ions available, less water will be able to enter ciliary epithelial cells [10]. The second generation carbonic anhydrase inhibitors include Dorzolamide, brand name Trusopt, and Brinzolamide, brand name Azopt. These are topical therapeutics and are able to reduce intraocular pressure 15–20% from initial value [10]. Common side effects to second generation carbonic anhydrase inhibitors include allergic reaction, ocular burning, ocular stinging, bitter taste, superficial punctate keratitis, blurred vision, headache, and dizziness [10]. Additionally, there is evidence that topical carbonic anhydrase inhibitors enhance ocular blood flow in the retina and optic nerve with a potentially favorable response on retinal ganglion cells [9]. Dorzolamide and Brinzolamide are effective IOP lowering agents and should be used as second line treatments and in combination with other glaucoma medications.

6.1 Nitric oxides

Although some may consider Latanoprostene bunod to be a subcategory of prostaglandin analogs, it should have its own category as it employs a different mechanism of action. This recently developed medication is commonly known by its brand name Vyzulta. It is essentially a modified prostaglandin analog that acts by dual mechanism. It is applied to the eye topically and is first hydrolyzed by endogenous esterases into Latanoprost acid, which is the active component of Latanoprost. What makes Latanoprostene bunod unique is that it also breaks down into butanediol mononitrate, which is further broken down into nitric oxide and inactive 1,4 butanediol [7]. The Latanoprost acid component increases aqueous outflow through the uveoscleral pathway, while the nitric oxide component induces relaxation within the trabecular meshwork. It works to increase aqueous outflow through the trabecular meshwork pathway and Schlemm's canal [7]. There have been several studies completed to examine safety and efficacy of latanoprostene Bunod. The Apollo Study concluded that latanoprostene bunod demonstrated significantly greater intraocular pressure lowering than Timolol, another commonly used IOP lowering agent. The study shows that latanoprostene bunod was effective and safe in adults with primary open angle glaucoma and ocular hypertension over 3 months of treatment [13]. Another study compared the efficacy of Latanoprostene Bunod to Latanoprost. They were able to conclude that Latanoprostene Bunod was significantly more effective than Latanoprost at lowering IOP with similar side effects [14]. This is a significant study because Latanoprost is widely accepted as the clinical benchmark of glaucoma treatment to which other drugs should be compared.

6.2 Prostaglandin analogs

Prostaglandin analogs are the most widely used pharmacologic treatment for primary open angle glaucoma and for good reason. They have been the clinical gold standard of IOP lowering agents since the 1990s. They are extremely efficacious, present very little risk, and are easy to use with a dosage schedule of once a day before bed. It is important to review the physiology of prostaglandins to understand their mechanism of action. Prostaglandins are proinflammatory molecules produced when arachidonic acid is metabolized by cyclooxygenase enzymes, COX-1 and COX-2 [15]. Basal levels are produced by COX-1, and further increased by COX-2 [15]. The 5 classes of prostaglandins are E2, F2 (PGF2), I2, D2, and Thromboxane A [16]. Each of these molecules is able to elicit different responses when interacting with their corresponding G-protein coupled receptor (GPCR). FP receptor proteins in humans have been detected throughout the anatomy of the eye. They are located in the corneal epithelium, ciliary epithelium, the circular portion of the ciliary muscle, and iris stromal and smooth muscle cells [15]. FP receptors activate metabolism through G-coupled proteins, resulting in the increase of intracellular calcium concentrations and modulation of various signaling cascades [15]. Prostaglandin analog mechanism of action works through decreasing IOP by increasing uveoscleral outflow. They activate prostaglandin receptors in ciliary muscle, iris root, and sclera to effectively induce relaxation of ciliary muscle and alter cytoskeletal remodeling of the extracellular matrix of the uveoscleral pathway. They also may enhance aqueous outflow via FP receptors present in the trabecular meshwork by lowering resistance [15]. Binding to prostaglandin receptors in the ciliary muscle allows for widening and decompression of the fluid filled spaces along the ciliary muscle bundles [10]. Simply put, prostaglandin F2 analogs bind to prostaglandin receptors to promote muscle relaxation and extracellular matrix remodeling in the ciliary muscle and trabecular meshwork, increasing aqueous humor outflow. These mechanisms provide prostaglandin F2 receptor agonists like Latanoprost, Bimatoprost, and Travoprost with the ability to be very effective IOP reducing agents.

Prostaglandin analogs are the most effective known pharmacologic treatment in terms of IOP reduction for primary open angle glaucoma [7]. They are able to reduce IOP by as much as 25–33% [17]. They rose to prominence after FDA approval in the 1990s when their IOP lowering success and lack of serious adverse effects became widely known. They were able to rapidly replace beta blockers as the gold standard of glaucoma pharmacologic therapy. The brand name version of Latanoprost, Xalatan became the first drug to exceed 1 billion US dollars in annual sales [9]. Although valued for their lack of serious side effects, some adverse reactions are noted. These include conjunctival hyperemia, burning, stinging, eyelash growth and hyperpigmentation, increased periocular skin pigmentation, increased iris pigmentation and loss of periorbital fat. Additionally, some cases of cystoid macular edema as well as reactivation of herpes keratitis and anterior uveitis have been documented [7, 10]. The incidence of increased iris pigmentation was found to be higher than expected. One observational cohort study concluded that 69.7% of patients developed iridial anisochromia after chronic Latanoprost use [18]. Therefore, it is suggested to use Latanoprost bilaterally in order to avoid hyperpigmentation on one side of the face. However, a lack of serious systemic side effects keeps Latanoprost as the first line glaucoma medication.

Several clinical studies have examined the efficacy of prostaglandin analogs against other members of this class as well as against other intraocular pressure reducing agents. When comparing the IOP reducing effectiveness of prostaglandin analogs, Bimatoprost, brand name.

Lumigan, consistently ranks the highest. This is usually followed by Latanoprost and Travoprost.

One meta-analysis of 17 different clinical studies suggests that Bimatoprost is the most effective of these three commonly used prostaglandin analogs following long term treatment, but also appears to have the most adverse effects and lower ocular tolerability [19]. Reviews suggest similar findings, with the efficacy of Bimatoprost to be about 1 mm Hg reduction superior to Latanoprost, yet Bimatoprost having the least appealing adverse effect profile [9]. Although clearly deserving of medical use, Bimatoprost's risk of lower ocular tolerability compared to Latanoprost puts Latanoprost ahead in terms of optimal clinical efficiency. Numerous additional studies prove prostaglandin analogs to be the most potent intraocular pressure reducing agents. One analysis displays the results of 50 studies with over 9000 patients included. Classes of IOP reduction used in the comparison includes beta blockers, alpha adrenergic agonists, and carbonic anhydrase inhibitors. It indicates that of all monotherapies examined, prostaglandin monotherapy showed the largest IOP reducing effect [20]. Another study examined the mean reductions in IOP after 3 months of usage. Bimatoprost averaged 5.61, Latanoprost 4.85, Travoprost 4.83, Timolol 3.70, Brimonidine 3.59, and Dorzolamide 2.59 mm Hg among others [21]. Prostaglandin analogs and specifically Latanoprost with its higher degree of ocular tolerability are well studied, potent, IOP reducing agents for glaucoma treatment. As the primary option for clinical use since the 1990s, new glaucoma therapies should continue to be compared against them.

7. Rho-kinase inhibitors

Rho-kinase inhibitors, also known as ROCK inhibitors are a relatively new class of antiglaucoma therapeutics. Netarsudil, brand name Rhopressa, was approved by the US Food and Drug Administration to be used for the treatment of glaucoma in 2017 [9]. Netarsudil is a an effective new therapy and is notable for its IOP lowering effects through a variety of different mechanisms. ROCK inhibitors work to decrease intraocular pressure (IOP) by inhibiting ROCK, a ubiquitous downstream effector protein that regulates the cell cytoskeleton [22]. ROCK consist of an amino-terminal serine-threonine kinase domain that is followed by a coiled-coilforming region and other functional motifs at a carboxyl terminus [22]. The carboxyl terminal domain forms an autoinhibitory loop that folds back onto the kinase domain and inhibits its activity [22]. The natural function of ROCK is to promote the assembly of actin stress fibers and focal adhesions within the trabecular meshwork [23]. As a potent rho-kinase inhibitor and amino.

Isoquinaline amide, Netarsudil also inhibits norepinephrine transporter (NET) [23]. There have been many animal studies that examine the effects of rho-kinase inhibitors. One study concluded that 2 ROCK inhibitors were able to relax ciliary arteries in rabbits to improve blood flow and suggested that they may have the potential to increase optic nerve blood flow to treat glaucoma [24]. Netarsudil is unique in that it has three proven mechanisms of action by which it reduces intraocular pressure in humans. It relaxes the trabecular meshwork, Schlemm's canal, and ciliary muscle to increase aqueous humor outflow through the conventional trabecular pathway [22, 23]. It acts by decreasing actomyosin-driven cellular contraction and reducing production of fibrogenic extracellular matrix proteins [25]. It also reduces IOP by decreasing the production of aqueous humor as well as lowering episcleral venous pressure [23]. These different mechanisms make Netarsudil an effective pharmacologic glaucoma treatment, however more research is needed to further study its additional effects on animals in human trials.

Several clinical studies have occurred to examine the safety and efficacy of chronic Netarsudil use. In a 12 month long clinical trial of 756 eligible patients, the ROCKET-2 study group showed that the most frequently reported adverse effects associated with once daily Netarsudil use were ocular. The most common effect was hyperemia, followed by corneal verticillata and conjunctival hemorrhage [25]. All of these effects were reported as mild and there was no clinically meaningful impact of corneal verticillata on visual function upon observational follow up [25]. Other reported adverse effects of Netarsudil include instillation site pain, blurred vision, increased lacrimation, eye pruritus, and erythema of the eyelid [17]. Phase 3 clinical trials ROCKET-1 and ROCKET-2 reported that the Netarsudil was able to elicit statistically significant reductions from baseline intraocular pressure and was evaluated to be noninferior to the IOP reducing effects of Timolol on open angle glaucoma and ocular hypertension patients [26]. The mean decrease from baseline IOP for once daily Netarsudil ranged from 3.3–4.6 mm Hg while those for Timolol ranged from 3.7–5.1 mm Hg [26]. This provided a 16–21% decrease in intraocular pressure for Netarsudil and a 18–23% decrease in IOP with Timolol [26].

To truly evaluate the potential of Netarsudil to become a popular clinical treatment, it is important to compare it against the efficacy of Latanoprost. One clinical trial comparing Netarsudil to Latanoprost concluded that Netarsudil was approximately 1 mm Hg less effective than Latanoprost in patients with unmedicated intraocular pressures between 22 and 35 mm Hg [27].

Interestingly, they also remarked that the adverse effect of ocular hyperemia was more prominent in both tested concentrations of Netarsudil than they were for Latanoprost [27]. Additionally, many large scale literature reviews and meta-analyses have been conducted comparing the older but remarkably effective Latanoprost to the newly approved formulation of netarsudil. Results are generally consistent in suggesting that Netarsudil is inferior to Latanoprost in reducing intraocular pressure in primary open angle glaucoma and ocular hypertension patients. They also conclude that Netarsudil has the potential for more ocular adverse effects [28, 29]. Although consistently proving slightly inferior to Latanoprost and similar in efficacy to Timolol, netarsudil usage should continue to grow as a secondary line of defense against glaucoma. The unique 3 part mechanism by which it reduces intraocular pressure could be useful in treating secondary glaucoma or in patients who do not respond well to more established IOP reducing agents. With further research, clinical trials, and investigation into combination therapy, Netarsudil has the potential to become a frequently utilized pharmacologic glaucoma treatment.

8. Combination therapy

Fixed dose combinations (FDCs) are an extremely useful tool in treating primary open angle glaucoma. The use of multiple pharmacologic agents to treat glaucoma via different mechanistic pathways has long been common clinical practice. However, the development and subsequent use of fixed dose combinations often presents an increased efficacy to the components of the combination when used alone.

Additionally, FDC's have the potential to facilitate greater adherence to medical recommendations. Reducing the number and cost of topical drops for a patient will lead to easier instructions, happier patients, and better clinical outcomes. This is especially true in medically underserved communities. As discussed previously, glaucoma disproportionately affects Black and Latino populations, and these populations have been shown to be affected by higher rates of inconsistent follow up visits [30]. With better access to, and more efficacious fixed dose combinations, pharmacologic glaucoma therapy has the potential to increase its efficiency and improve patient outcomes.

Modern fixed dose combinations frequently used in the United States today include Dorzolamide-Timolol (Cosopt), Brimonidine-Timolol (Combigan), and Brinzolamide-Brimonidine (Simbrinza). These combinations are all efficacious in their IOP lowering potential. In one study comparing combination therapies, those with a prostaglandin analog demonstrated more powerful IOP lowering efficacy [20]. Consequently, the recently FDA approved fixed dose combination of Netarsudil-Latanoprost (Rocklatan), has been shown to be a potent anti-glaucoma medication. This is significant in that it combines the widely accepted prostaglandin analog that is known as the most efficacious IOP lowering agent, with a new drug with a unique and effective mechanism. In a phase 3 clinical trial comparing the safety and efficacy of Netarsudil Latanoprost with its individual components, it consistently outperformed both Netarsudil and Latanoprost monotherapy [31]. Netarsudil-Latanoprost FDC was superior to its individual components at each time point measured in the study. It lowered IOP by an additional 1.8–3.0 mm Hg vs. Netarsudil and 1.3–2.5 mm Hg vs. Latanoprost [31]. After 3 months of treatment, the proportion of patients achieving a mean diurnal IOP of less than or equal to 15 mm Hg was 43.5% for the FDC, 22.7% for Netarsudil, and 24.7% for Latanoprost [31]. Conjunctival hyperemia was the highest reported adverse effect and the range of effects was comparable to Netarsudil monotherapy. The FDC did however produce a higher percentage of conjunctival hyperemia with 53.4% patients compared to 41% with Netarsudil and 14% with Latanoprost [31].

Severity of effects however was reported as mild and it was concluded that the FDC of Netarsudil-Latanoprost was safe and effective for use. Pooled data by the researchers of the different phase 3 clinical trials confirms prior results concluding that once daily Netarsudil Latanoprost FDC produced statistically significant and clinically relevant reductions in mean IOP when compared to its individual components [32]. Additionally, a systematic review of clinical trials comparing Netarsudil-Latanoprost FDC to each component yielded similar results. It concluded that the FDC was superior in IOP lowering effectiveness, but with some concern over the higher incidence of adverse effects when compared with Latanoprost specifically [33]. With significant efficacy and the ease of once daily dosing, Netarsudil-Latanoprost has the potential to become a staple of combination glaucoma treatment.

9. Oral

The first generation of carbonic anhydrase inhibitors includes Acetazolamide, brand name Diamox. It is a potent IOP reducing agent with a 25–30% reduction possible [17]. Acetazolamide is different from other glaucoma medications in that it is often used in a pill form. It has been used as a systemic therapeutic to treat glaucoma for over 50 years, but presents significant adverse effects due to its inhibition of carbonic anhydrase in tissues other than the eye [17]. Therefore it is not recommended for chronic use and is best utilized as a temporary IOP reducing measure in instances of severe elevation. Medication in pregnancy is usually the second treatment option. The FDA classifies medication into pregnancy risk categories. Category A (no risk) to Category E (the highest risk). There are no class A glaucoma drugs on the market. Class B includes alpha adrenergic agonists (Brimonidine, Apraclonidine) and nonselective alpha and beta agonist (epinephrine). Class C drugs include beta blockers, carbonic anhydrase inhibitors and prostaglandin analogs. Netarsudil and Latanaprostene Bunod have yet to be classified. A great alternative is SLT therapy. Medication in children is usually safe, however caution should be taken to avoid the alpha adrenergic agonist brimonidine as it can lead to respiratory and CNS depression (lower level of consciousness and hypotonic). Medication in the elderly is safe as long as a thorough history is taken and all comorbidities are taken into consideration when prescribing any medication. As there are increased risks of underlying medical problems such as heart, lung, and kidney disease, any prescription given should be prescribed cautiously. Generic medications are usually safe and many insurance companies will only prescribe them. They have the same active ingredients and effect as brand name medications, but may be a different, size, color, or shape. They are usually less expensive than brand name medications. Many drug companies may make versions of the same medication. The FDA requires that the same active ingredients in brand name drugs are in generic drugs.

10. Natural therapies

Glaucoma management focuses on reducing intraocular pressure to halt visual field loss. Ophthalmologists are often required to invest massive amounts of time and energy when managing glaucoma cases by working to maintain IOP in a homeostatic range in order to preserve the anatomical integrity of the optic nerve [34]. In order to prevent this neuropathy from being expressed, vigorous medical interventions are required, and the treatment modalities should be personalized given the fact that no two individual cases of glaucoma are identical to each other [34]. Data indicates that by lowering IOP, there can be a beneficial impact on the slowing of glaucomatous progression [35]. The following natural substances have been identified in our review process as potential treatments to help reduce IOP in glaucoma patients while having limited to no adverse effects.

10.1 Persimmon leaves

Persimmon (*Diospyros kaki*) is native to Eastern Asia, including Korea, China and Japan [36]. It is plentiful in carotenoids (e.g., lutein and zeaxanthin), which can protect eyes from detrimental optic disorders, such as glaucoma, cataracts, and macular degeneration [36]. Just like its fruit counterpart, the persimmon leaves house copious amounts of biologically active compounds, including organic acids, polyphenols, flavonoids, and vitamins, most of which are known to exert beneficial treatment properties, such as aggressive radical-scavenging, antioxidant properties, and immune-enhancing traits [36]. In one particular study, Ahn et al. efficaciously created a microbeads-induced ocular hypertension glaucoma model and compared the results with those of DBA/2 mice, as both animal models develop age- dependent glaucoma phenotype. Furthermore, both animal models can effectively raise and sustain IOP, which is optimal for validating the pharmacologic efficacy of the persimmon leave extract. Undeniably, IOP in both the DBA/2 and microbeads-induced ocular hypertension mouse models was raised to a level substantially higher than that of the control group mice. Persimmon leave extract distribution via ocular eye drop absorption produced IOP values akin to those recorded from treatment with prostaglandin analogs (i.e., Xalatan), which is the most commonly used pharmacologic agent used in glaucoma management [37].

10.2 Omega-3-fatty acids

One specific dietary factor that has piqued the interest of researchers has been whether or not omega-3 essential fatty acids can prevent and/or support therapies for a host of prevalent health conditions, including glaucoma [38]. As omega-3 and omega-6 fatty acids compete in vivo for enzymes regulating their metabolism, the ratio of consumed omega-3 to omega-6 essential fatty acids regulates the inflammatory condition of the body, with omega-3's amplifying prostaglandin metabolism which subsequently promotes the production of anti-inflammatory eicosanoids [38]. This chain reaction helps keep inflammation in the body to a minimum [38]. Data indicates that individuals who ingest high amounts of omega-3 essential fatty acids in their diet tend to express a reduced risk of heart disease mortality, abated age-related neurological decline, and a decreased risk of age-related macular degeneration. Additionally, an absence of omega-3 essential fatty acids may expose individuals to optic diseases in the elderly stages of life [38]. It serves that the possible perk of omega-3 fatty acid supplementation has undergone assessment in clinical trials for ocular diseases that show an increased prevalence with age, particularly age-related macular degeneration and glaucoma [39]. Data from the cross-sectional National Health and Nutrition Examination Survey indicated that elevated amounts of daily consumption of the long-chain, polyunsaturated omega-3 fatty acids, eicosapentaenoic acid, and docosahexaenoic acid, was associated with a lesser prevalence of glaucomatous optic neuropathy [39]. Moreover, in that same cross-sectional study, it was shown that the likelihood of an individual having glaucoma was nearly three times as high in individuals whose daily dietary total long-chain omega-3 consumption level was in the second and third quartiles, compared to those in the first quartile [39].

10.3 Ginseng

Ginseng, (i.e., the roots of *Panax ginseng*, *P. notoginseng*, and *P. quinquefolius*), has been broadly utilized as a panacea for a wide spectrum of medical conditions for many years and is currently being scientifically analyzed for its efficacy in the treatment of certain ocular diseases, such as glaucoma [40]. Numerous studies have shown the effects of ginseng extract or ginsenoside on all four of the major eye conditions previously listed in this review, including glaucoma [41]. The ability to unearth natural treatments that would replace more toxic and expensive pharmaceutical agents to prevent or prolong glaucomatous progression and subsequent blindness would sustain a higher quality of life while also simultaneously easing the financial burden associated with managing this condition. Investigations have been initiated to reveal the preventative and therapeutic effects of ginseng and ginsenosides on a multitude of ocular diseases. There have been some strong indicators that suggest the biochemical mechanisms of antiinflammation, antioxidation, elimination of waste products, and inhibition of vascular endothelial growth factor are thought to be involved in the therapeutic outcomes for this natural substance [41].

10.4 Curcumin

Curcumin (diferuloylmethane) is a compound derived from the rhizome of the Indian spice turmeric (*Curcuma longa*) and belongs to the class of micronutrients known as the polyphenols [42]. Curcumin has been observed to play a role in slowing, and in some cases, even reversing age-related macular degeneration, diabetic retinopathy, retinitis pigmentosa, proliferative vitreoretinopathy, and retinal cancers [43]. This makes curcumin a particularly interesting agent when examining natural treatment modalities for the treatment of retinal disorders and glaucoma.

10.5 Baicalein

Baicalein is categorized as a flavonoid glycoside. After it loses a water molecule via hydrolysis, it converts into a closely related aglycone known as Baicalein [44]. Baicalein is commonly found in the plants of the genus Scutellaria (Lamiaceae) as a major component of leaves, root bark, and fruit while Baicalin is found in bountiful amounts in stem bark and leaves [44]. There have been multiple investigations that have concluded that both these molecules have a tremendous capacity to function as powerful anti-inflammatory mediators [44]. One of the chief anti-inflammatory mechanisms that these molecules use is reducing oxidative stress in the cellular environment, which in turn improves the antioxidant status of the host cell [45]. Additionally, we encountered data demonstrating that baicalein suppresses the net chloride-transport and fluid movement across the excised ciliary epithelium, potentially reducing the aqueous humor formation and IOP [46].

10.6 Saffron

Saffron (*Crocus sativus* L.) is one of the oldest cultural and culinary spices used. Saffron ingestion has been shown to catalyze significant radical scavenging, anti-inflammatory, and anti-apoptotic activities in multiple investigations [47]. Biochemical pathways are activated from the diverse collection of bioactive agents housed in the spice. In one study, a total of 34 randomized subjects including 17 patients (7 female and 10 male individuals) were selected to be in the experimental saffron group and 17 other patients (6 female and 11 male individuals) were slotted to be in the placebo group [48]. The average baseline IOP measured 12.9 ± 3.7 versus 14.0 ± 2.5 mmHg in the saffron and control groups, respectively ($p = 0.31$) [48]. After 21 days of treatment, IOP was significantly decreased to 10.9 ± 3.3 mmHg in the saffron group as compared to 13.5 ± 2.3 mmHg in the control group ($p = 0.013$) [48]. Collectively, this set of data points strongly suggests that oral aqueous saffron extract seems to exert an ocular hypotensive effect in primary open-angle glaucoma patients.

10.7 Dietary nitrates

There is a consensus that dietary nitrates are fundamentally inert and are only activated biochemically after reduction to nitrite [49]. Nitrate serves as a source, via successive reduction, to produce nitrite and nitric oxide as well as a host of other metabolic compounds. In one specific study, the authors set out to find evidence that supports the association between dietary nitrate intake, derived from green leafy vegetables, and the onset of glaucoma. The key finding that arose from the data was that compared with the lowest quintile of dietary nitrate intake (80 mg/d), the highest quintile (240 mg/d) was associated with a 21% lower risk of all glaucoma and 44% lower risk of glaucoma with early paracentral visual field loss [50]. This key finding could have a significant impact on the future direction of treatment modalities for glaucoma if the association of higher dietary nitrate consumption is correlated with a lower glaucoma risk and this relationship is confirmed via observational or interventional studies [50].

10.8 *Ginkgo Biloba*

In modern times, *Ginkgo biloba* is one of the most prescribed and used herbal medicines in the United States with market sales of over $150 million in 2015 [51]. As far as a potential therapeutic agent for the management of glaucoma, there are three main properties of this herbal extract that make it appealing to researchers and ophthalmologists: (1) ability to increase vascular flow and reduce blood viscosity; (2) antioxidative inducing effects; and (3) neuroprotective enhancing properties [52].

10.9 Green tea

Green tea is an ancient Chinese drink derived from the *Camellia sinensis* plant. Green tea has been studied intensely for its possible protective effects against heart disease and cancer and this discussion is designed to probe its possible role in the treatment and management of glaucoma. We encountered a significant study which set out to show the therapeutic effect of green tea extract on ischemia-induced retinal ganglion cell degeneration in an animal model of rats. There were six fundamental takeaways that were reported to be significant out of this animal model study and they are as follows: (1) oral administration of 275 mg/kg of green tea extract was safe and no toxic or detrimental effects were observed in the retinas of the observed rats; (2) ischemic reperfusion induced retinal ganglion cell degeneration in rats; (3) ischemia injury induces a pathological cocktail of inflammation, oxidative stress, and cell apoptosis in the retina; (4) green tea extract treatment effectively ameliorates ischemia-induced retinal ganglion cell degeneration; (5) green tea extract treatment inhibited the pupillary light reflex and retinal ganglion cell impairment in rats Naturak Medication Options after induced ischemic injury; and (6) green tea extract treatment reduced cell apoptosis, oxidative stress, and inflammation and increased survival signal in the retina [53]. Together, these revelations highlight a significant protective effect of green tea extract on retinal ganglion cell degeneration caused by ischemia and suggest a beneficial therapeutic application of green tea extract for the management of glaucoma.

10.10 Hesperidin

Hesperidin is a major flavonoid found in almost all citrus fruit and various polyherbal formulations and is one of the chief nutrients of the vitamin P class [54]. Virtually all observed therapeutic mechanisms are associated with hesperidin's ability to promote antioxidant defense mechanisms while concurrently suppressing the synthesis of pro inflammatory cytokines [55]. Flavonoids such as hesperidin, present a strong case for use as a therapeutic agent for glaucoma management due to their lower toxicity at higher doses and prolonged period of treatment duration that has been shown in animal model studies [56]. In one study, Lu et al. selected hesperidin due to its strong antioxidant potential. Kara et al. have reported the therapeutic effect of hesperetin against apoptosis in ischemia-induced retinal injury model of rats [56]. Moreover, Maekawa et al. have described the neuroprotective effect of hesperidin in N-methyl-D-aspartate -induced retinal injury [56]. There have been multiple experiments that have highlighted the therapeutic potential of bioflavonoids against ocular disorders. Estruel-Amades et al. have noted the therapeutic potential of hesperidin against oxidative stress in rats [56]. This data indicates that hesperidin supplementation was effective against glaucoma in experimental rat models.

11. Lifestyle modification

As with most chronic conditions, the expression of different glaucoma prevalence rates within different subsets of our populations is due to a delicate balance of both genetic and environmental factors being co-expressed simultaneously. Therefore, no review on natural treatments for the management of glaucoma would be complete without addressing the fundamental areas of key lifestyle modifications that can be implemented to improve upon the prevention and treatment regrading desirable health outcomes of glaucoma cases. There have been a multitude of active chemical compounds identified in tobacco smoke that have been found to have toxic ocular effects via oxidative pathways or ischemic mechanisms [57]. We identified a recent study derived from a population sample drawn from the United States National Health and Nutrition Examination Survey that identified the correlation that high volume smoking (i.e., multiple packs of cigarettes per day) was associated with greater risk of the onset of glaucoma with a specific odds ratio being calculated at 1.7 [58].

One of the most studied lifestyle modifications is exercise. It has been demonstrated that aerobic inducing exercises can lead to decreased IOP, while activities such as yoga and strength training (i.e., isometric exercise) can have the inverse effect of raising IOP levels by increasing intrathoracic pressure [59]. When specifically examining the association between glaucoma prevalence rates and exercise participation, there was an influential population-based study that highlighted a U-shaped association for male participants in the study [60]. This nonlinear data showcased the association of high and low levels of exercise intensity with higher prevalence rates of glaucoma when evaluated against data produced from moderate intensity exercise sessions [60]. Therefore, it seems reasonable to state that exercise, when performed ideally at a rate of three times a week can be suggested to improve a patient's general health and decrease risk of glaucoma.

12. Racial and ethnic disparities in glaucoma

Disparities in healthcare exist across a wide range of diseases and medical conditions. It is important to acknowledge racial and ethnic disparities when reviewing a specific disease in order to promote awareness and improve patient outcomes. Members of minority groups in the.

United States and the world often have less access to and receive lower quality medical care than White individuals. They also have lower rates of representation in clinical studies than White Americans. This presents a significant problem in the treatment of glaucoma. The standard treatment for primary open angle glaucoma is reducing intraocular pressure. Therefore, it is critically important to adhere to medical and surgical appointments and to follow therapeutic regiments in order to increase clinical outcomes. Black and Latino racial and ethnic patient status has been shown to be an independent risk factor for inconsistent follow up visits at appointments and lower rates of glaucoma testing. Cost, health literacy, and access to screening are some of the greatest obstacles that Black and Latino patients face with respect to glaucoma [30, 61]. Another factor that could add to this disparity is that less than 2% of ophthalmologists in the United States identify as Black or Latino [30]. This has the potential to negatively affect patient interactions, trust, communication, and satisfaction.

13. Disparities in care: clinical trials and beyond

Globally, it is estimated that 76 million people suffer from glaucoma with a projected jump to 112 million by 2040 [2]. This number indicates the importance of studying this disease and continuing to search for therapeutics that can improve patient outcomes. This involves improving racial and ethnic diversity in study participants. Glaucoma has been shown to affect Black and Latino populations at a significantly higher rate than White populations in the United States. It is reported that this disease is 7 times more likely to cause blindness in Black individuals when compared with White individuals, and 15 times more likely to cause visual impairment in Black individuals than White individuals [62]. As mentioned previously, primary open angle glaucoma (POAG) is the most common form of glaucoma. Additional statistics indicate that Black individuals had a higher prevalence rate of POAG than White individuals in the United States. Black persons had the highest rate at 3.4% while White persons had a rate of 1.7% [62]. Additionally, increasing the diversity of clinical study participants will help with understanding how to better improve clinical outcomes for Black and Latino populations. Allison et al. reported that between 1994 and 2019, White participants made up 70.7% of study populations, while Black participants made up only 16.8% and Latino individuals only 3.4% [62]. It has been theorized that there could be genetic variants associated with the increased rate of glaucoma in Black populations. However, previous observational studies have proven to be inconclusive [62]. It is commonly agreed upon that further research is needed to study the genetic mechanisms of primary open angle glaucoma.

Although more genetic studies into POAG disparities in race and ethnicity need to be performed, it is clear that that socioeconomic status plays a central role in these epidemiologic differences. Socioeconomic status is a social determinant of health and it is commonly understood that racial and ethnic minority groups in the United States experience more socioeconomic disadvantages when compared to White individuals [62]. As socioeconomic disadvantages increase, the use of eye care services for age related diseases decreases [63]. Minority populations in the United States are underserved and over-affected by POAG.

14. Conclusion

Glaucoma is the leading cause of irreversible blindness worldwide. With a population of patients expected to grow to 112 million people globally by 2040 [2], it is important to continue to develop new pharmacologic therapies to control intraocular pressure. While the prostaglandin analog latanoprost continues to be the widely accepted first line IOP reducing agent used today, the recently developed Rho-kinase inhibitor netarsudil, and fixed dose combination of netarsudil-latanoprost should continue to increase in utilization as both second line and combination therapy drugs. In addition to once daily dosing and complementary mechanisms of action, future directions of glaucoma treatment should work to increase clinical outcomes through more research on natural therapeutics, sustained release mechanisms, and achieving longer acting effects. Prioritizing patient care by decreasing medication burden will lead to increased compliance, improved satisfaction, and better patient outcomes.

Author details

Karen Allison[1,2*], Kevin Morabito Jr[1], Deepkumar Patel[3,4] and Brandon W. Montoya[5]

1 Flaum Eye Institute, University of Rochester, Rochester, NY, USA

2 Prevention of Blindness from Glaucoma and Macular Degeneration, New York Medical College School of Public Health, Floral Park, NY, USA

3 Public Health, University of California, San Francisco, CA, USA

4 Prevention of Blindness from Glaucoma and Macular Degeneration, Floral Park, NY, USA

5 Prevention of Blindness from Glaucoma and Macular Degeneration, Public Health New York Medical College, Floral Park, NY, USA

*Address all correspondence to: kallisonmdpc@gmail.com

References

[1] Jonas JB et al. Glaucoma. Lancet. 2017;**390**(10108):2183-2193

[2] Stein JD, Khawaja AP, Weizer JS. Glaucoma in adults—Screening, diagnosis, and management: A review. JAMA. 2021;**325**(2):164-174

[3] Kass MA et al. The ocular hypertension treatment study: A randomized trial determines that topical ocular hypotensive medication delays or prevents the onset of primary open-angle glaucoma. Archives of Ophthalmology. 2002;**120**(6):701-713. discussion 829-30

[4] Garway-Heath DF et al. Latanoprost for open-angle glaucoma (UKGTS): A randomised, multicentre, placebo-controlled trial. Lancet. 2015;**385**(9975):1295-1304

[5] Heijl A et al. Reduction of intraocular pressure and glaucoma progression: Results from the early manifest glaucoma trial. Archives of Ophthalmology. 2002;**120**(10):1268-1279

[6] Leske MC et al. Factors for glaucoma progression and the effect of treatment: The early manifest glaucoma trial. Archives of Ophthalmology. 2003;**121**(1):48-56

[7] Garg A, Gazzard G. Treatment choices for newly diagnosed primary open angle and ocular hypertension patients. Eye (London, England). 2020;**34**(1):60-71

[8] Weinreb RN, Aung T, Medeiros FA. The pathophysiology and treatment of glaucoma: A review. JAMA. 2014;**311**(18):1901-1911

[9] Casson RJ. Medical therapy for glaucoma: A review. Clinical & Experimental Ophthalmology. 2022;**50**(2):198-212

[10] Schmidl D et al. Pharmacotherapy of glaucoma. Journal of Ocular Pharmacology and Therapeutics. 2015;**31**(2):63-77

[11] Toris CB, Camras CB, Yablonski ME. Acute versus chronic effects of brimonidine on aqueous humor dynamics in ocular hypertensive patients. American Journal of Ophthalmology. 1999;**128**(1):8-14

[12] Seki M et al. Topically administered timolol and dorzolamide reduce intraocular pressure and protect retinal ganglion cells in a rat experimental glaucoma model. The British Journal of Ophthalmology. 2005;**89**(4):504-507

[13] Weinreb RN et al. Latanoprostene bunod 0.024% versus timolol maleate 0.5% in subjects with open-angle glaucoma or ocular hypertension: The Apollo study. Ophthalmology. 2016;**123**(5):965-973

[14] Weinreb RN et al. A randomised, controlled comparison of latanoprostene bunod and latanoprost 0.005% in the treatment of ocular hypertension and open angle glaucoma: The VOYAGER study. The British Journal of Ophthalmology. 2015;**99**(6):738-745

[15] Zhou L, Zhan W, Wei X. Clinical pharmacology and pharmacogenetics of prostaglandin analogues in glaucoma. Frontiers in Pharmacology. 2022;**13**:1015338

[16] Doucette LP, Walter MA. Prostaglandins in the eye: Function, expression, and roles in glaucoma. Ophthalmic Genetics. 2017;**38**(2):108-116

[17] Wang T et al. Topical medication therapy for glaucoma and ocular hypertension. Frontiers in Pharmacology. 2021;**12**:749858

[18] Teus MA, Arranz-Márquez E, Lucea-Suescun P. Incidence of iris colour change in latanoprost treated eyes. The British Journal of Ophthalmology. 2002;**86**(10):1085-1088

[19] Tang W et al. Efficacy and safety of prostaglandin analogues in primary open-angle glaucoma or ocular hypertension patients: A meta-analysis. Medicine (Baltimore). 2019;**98**(30):e16597

[20] Li F, Huang W, Zhang X. Efficacy and safety of different regimens for primary open-angle glaucoma or ocular hypertension: A systematic review and network meta-analysis. Acta Ophthalmologica. 2018;**96**(3):e277-e284

[21] Li T et al. Comparative effectiveness of first-line medications for primary open-angle glaucoma: A systematic review and network meta-analysis. Ophthalmology. 2016;**123**(1):129-140

[22] Al-Humimat G et al. Investigational rho kinase inhibitors for the treatment of glaucoma. Journal of Experimental Pharmacology. 2021;**13**:197-212

[23] Dasso L et al. Profile of netarsudil ophthalmic solution and its potential in the treatment of open-angle glaucoma: Evidence to date. Clinical Ophthalmology. 2018;**12**:1939-1944

[24] Watabe H, Abe S, Yoshitomi T. Effects of rho-associated protein kinase inhibitors Y-27632 and Y-39983 on isolated rabbit ciliary arteries. Japanese Journal of Ophthalmology. 2011;**55**(4):411-417

[25] Kahook MY et al. Long-term safety and ocular hypotensive efficacy evaluation of netarsudil ophthalmic solution: Rho kinase elevated IOP treatment trial (ROCKET-2). American Journal of Ophthalmology. 2019;**200**:130-137

[26] Serle JB et al. Two phase 3 clinical trials comparing the safety and efficacy of netarsudil to timolol in patients with elevated intraocular pressure: Rho kinase elevated IOP treatment trial 1 and 2 (ROCKET-1 and ROCKET-2). American Journal of Ophthalmology. 2018;**186**:116-127

[27] Bacharach J et al. Double-masked, randomized, dose-response study of AR-13324 versus latanoprost in patients with elevated intraocular pressure. Ophthalmology. 2015;**122**(2):302-307

[28] Clement Freiberg J et al. Rho kinase inhibitor for primary open-angle glaucoma and ocular hypertension. Cochrane Database of Systematic Reviews. 2022;**6**(6):Cd013817

[29] Gonzalez LE, Boylan PM. Netarsudil for the treatment of open-angle glaucoma and ocular hypertension: A literature review. The Annals of Pharmacotherapy. 2021;**55**(8):1025-1036

[30] Siegfried CJ, Shui YB. Racial disparities in glaucoma: From epidemiology to pathophysiology. Missouri Medicine. 2022;**119**(1):49-54

[31] Asrani S et al. Netarsudil/Latanoprost fixed-dose combination for elevated intraocular pressure: Three-month data from a randomized phase 3 trial. American Journal of Ophthalmology. 2019;**207**:248-257

[32] Asrani S et al. Fixed-dose combination of netarsudil and latanoprost in ocular hypertension and open-angle glaucoma: Pooled efficacy/safety analysis of phase 3 MERCURY-1

and -2. Advances in Therapy. 2020;**37**(4):1620-1631

[33] Luo N et al. Efficacy and safety of netarsudil/latanoprost fixed-dose combination vs. monotherapy in open-angle glaucoma or ocular hypertension: A systematic review and meta-analysis of randomized controlled trials. Frontiers in Medicine (Lausanne). 2022;**9**:923308

[34] Davis BM et al. Glaucoma: The retina and beyond. Acta Neuropathologica. 2016;**132**(6):807-826

[35] Sena DF, Lindsley K. Neuroprotection for treatment of glaucoma in adults. Cochrane Database of Systematic Reviews. 2017;**1**(1):Cd006539

[36] Mares J. Lutein and zeaxanthin isomers in eye health and disease. Annual Review of Nutrition. 2016;**36**:571-602

[37] Ahn HR et al. The intraocular pressure-lowering effect of persimmon leaves (Diospyros kaki) in a mouse model of glaucoma. International Journal of Molecular Sciences. 2019;**20**(21):5268

[38] Downie LE, Vingrys AJ. Oral omega-3 supplementation lowers intraocular pressure in normotensive adults. Translational Vision Science & Technology. 2018;7(3):1

[39] Wang YE et al. Association of dietary fatty acid intake with glaucoma in the United States. JAMA Ophthalmology. 2018;**136**(2):141-147

[40] Yuk J, Patel DN, Isaac G, Smith K, Wrona M, Olivos HJ, et al. Chemical profiling of ginseng species and ginseng herbal products using UPLC/QTOF-MS. Journal of the Brazilian Chemical Society. 2016;**27**(8):1476-1483. DOI: 10.5935/0103-5053.20160189

[41] Kim J, Han SY, Min H. Ginseng for an eye: Effects of ginseng on ocular diseases. Journal of Ginseng Research. 2020;**44**(1):1-7

[42] Kocaadam B, Şanlier N. Curcumin, an active component of turmeric (Curcuma longa), and its effects on health. Critical Reviews in Food Science and Nutrition. 2017;**57**(13):2889-2895

[43] Peddada KV et al. Therapeutic potential of curcumin in major retinal pathologies. International Ophthalmology. 2019;**39**(3):725-734

[44] Dinda B et al. Therapeutic potentials of baicalin and its aglycone, baicalein against inflammatory disorders. European Journal of Medicinal Chemistry. 2017;**131**:68-80

[45] Farzaei MH et al. Targeting inflammation by flavonoids: Novel therapeutic strategy for metabolic disorders. International Journal of Molecular Sciences. 2019;**20**(19):4957

[46] Li HL. Baicalein lowers intraocular pressure and increases outflow facility in mouse eye. Investigative Ophthalmology and Visual Science (IOVS), ARVO Abstract. Jun 2015;**56**:4853

[47] Fernández-Albarral JA et al. Neuroprotective and anti-inflammatory effects of a hydrophilic saffron extract in a model of glaucoma. International Journal of Molecular Sciences. 2019;**20**(17):4110

[48] Jabbarpoor Bonyadi MH, Yazdani S, Saadat S. The ocular hypotensive effect of saffron extract in primary open angle glaucoma: A pilot study. BMC Complementary and Alternative Medicine. 2014;**14**:399

[49] Ma L et al. Nitrate and nitrite in health and disease. Aging and Disease. 2018;**9**(5):938-945

[50] Kang JH et al. Association of dietary nitrate intake with primary open-angle glaucoma: A prospective analysis from the nurses' health study and health professionals follow-up study. JAMA Ophthalmology. 2016;**134**(3):294-303

[51] Gafner S. Adulteration of Ginkgo biloba Leaf Extract. 2018. Available from: https://cspinet.org/sites/default/files/botanical-adulterants-bulletin.pdf

[52] Kang JM, Lin S. Ginkgo biloba and its potential role in glaucoma. Current Opinion in Ophthalmology. 2018;**29**(2):116-120

[53] Yang Y et al. Green tea extract ameliorates ischemia-induced retinal ganglion cell degeneration in rats. Oxidative Medicine and Cellular Longevity. 2019;**2019**:8407206

[54] Preedy VR, Watson RR. Handbook of Nutrition, Diet, and the Eye. 2nd ed. Academic Press; 11 Jun 2019. ISBN: 9780128152454. eBook ISBN: 9780128152461

[55] Hesperetin - an overview. 2014. Available from: www.sciencedirect.com; https://www.sciencedirect.com/topics/nursing-and-health-professions/hesperetin

[56] Lu B et al. Anti-glaucoma potential of hesperidin in experimental glaucoma induced rats. AMB Express. 2020;**10**(1):94

[57] Solberg Y, Rosner M, Belkin M. The association between cigarette smoking and ocular diseases. Survey of Ophthalmology. 1998;**42**(6):535-547

[58] Law SM et al. Cigarette smoking and glaucoma in the United States population. Eye (London, England). 2018;**32**(4):716-725

[59] Natsis K et al. Aerobic exercise and intraocular pressure in normotensive and glaucoma patients. BMC Ophthalmology. 2009;**9**:6

[60] Lin SC et al. The relation between exercise and glaucoma in a south Korean population-based sample. PLoS One. 2017;**12**(2):e0171441

[61] Pleet A et al. Risk factors associated with progression to blindness from primary open-angle glaucoma in an African-American population. Ophthalmic Epidemiology. 2016;**23**(4):248-256

[62] Allison K, Patel DG, Greene L. Racial and ethnic disparities in primary open-angle glaucoma clinical trials: A systematic review and meta-analysis. JAMA Network Open. 2021;**4**(5):e218348

[63] Zhang X et al. Socioeconomic disparity in use of eye care services among US adults with age-related eye diseases: National health interview survey, 2002 and 2008. JAMA Ophthalmology. 2013;**131**(9):1198-1206

Chapter 6

Modalities of Measuring Intraocular Pressure: Updates and Advances

Sohum Sheth, Kevin Peng, Ankit Shah and Mark Disclafani

Abstract

Accurate measurement of intraocular pressure (IOP) is a vital part of ocular hypertension management to prevent progression toward glaucoma. IOP remains as one of the only significantly treatable risk factors for glaucoma, thus illustrating the importance of tonometry. Our chapter intends to compare the various modalities of tonometry, including applanation, indentation, rebound, dynamic contour, and transpalpebral scleral palpation. Moreover, we will discuss advances that enable continuous 24-h IOP monitoring, including contact lens sensors and implantable microsensors and review implications for glaucoma diagnosis and management. We will consider aspects, such as mechanisms, accuracy and precision, ease of use, and possible limitations or complications of each modality.

Keywords: intraocular pressure, tonometry, applanation, noncontact, indentation, rebound, dynamic contour

1. Introduction

Intraocular pressure (IOP) remains at the core of the ophthalmologic physical exam. As a careful balance between production and outflow of aqueous humor, disruptions to the equilibrium can lead to many pathologies, such as retinal detachment, uveitis, and glaucoma. IOP remains an important method of assessing the severity and progression of glaucoma, as well as efficacy of glaucoma treatments. With an appreciation of the biology that underlies aqueous humor dynamics, several instruments have been developed to obtain IOP measurements. The accuracy and precision of IOP readings have significant clinical implications and must be considered in the context in which the reading was taken, including the methodology used to obtain it. From a single reading through Tono-Pen tonometry to surgically implanted suprachoroidal microsensors with continuous IOP monitoring, each modality carries its own set of strengths and weaknesses. In this chapter, the various modalities to measure IOP are reviewed to provide the clinician with an understanding of the principles that enable IOP measurement and evidence regarding instrument use. In addition to evaluating conventional techniques of IOP measurements, an assessment of emerging techniques with the potential to revolutionize IOP monitoring and glaucoma management is discussed.

2. IOP principles: A function of aqueous humor production and outflow

IOP represents the magnitude of force exerted by the aqueous humor (AH) on the inner surface of the anterior eye. This relationship is shown in the Goldmann equation, which states that IOP = (F/C) + P, where F is aqueous flow rate, C is aqueous outflow, and P is the episcleral venous pressure. In effect, IOP indicates the balance between AH production and exit. Dysfunction in the balance between AH production and drainage can lead to increased IOP and subsequent pathology. Thus, an appreciation of the anatomy and physiology that dictates these aspects of aqueous humor is critical in the understanding of principles underlying IOP measurement.

2.1 AH production

Aqueous humor is produced by the cells in the ciliary body. Specifically, the anterior-most portion of the ciliary body constitutes the pars plicata, which contains the fingerlike projections of the ciliary process. The ciliary processes, in comparison to other regions of the ciliary body, contain higher amounts of interdigitations, mitochondria, rough endoplasmic reticulum, and gap junctions consistent with the function of aqueous humor production.

The process of AH production is conventionally broken into three steps. First, blood flows into the ciliary processes. Next, the hydrostatic and oncotic pressure gradient between the blood flow and the ciliary interstitium enables ultrafiltration of the plasma into the interstitium. Finally, the ciliary epithelium actively transports plasma ultrafiltrate from the basal side to the apical membrane and into the posterior chamber of the eye.

The ciliary body receives both sympathetic and parasympathetic innervation. The parasympathetic fibers arise from the Edinger–Westphal nucleus and the pterygopalatine ganglion. The sympathetic fibers arise from cervical superior ganglion and the carotid plexus. Hydrodynamic studies [1] have shown that the rate of AH turnover is around 1.0–1.5% of the anterior chamber volume per minute. Moreover, diurnal variations in AH production create a pattern referred to as the circadian rhythm of AH flow, with flow typically highest (3.0 μl/min) in the morning and lowest (1.5 μl/min) at night [2]. The mechanism underlying this patten is not well studied, but the actions of epinephrine on the ciliary epithelium are thought to mediate such effects [3].

2.2 Aqueous humor outflow

After production in the ciliary body, AH traverses from the posterior chamber to the anterior chamber around the lens and through the pupil (**Figure 1**). Two pathways—the conventional and nonconventional outflow routes—comprise AH exit from the anterior chamber. In the conventional pathway, AH traverses the trabecular meshwork to enter Schlemm's canal. From Schlemm's canal, AH enters the collector channels to join the episcleral venous system and, ultimately, systemic circulation. The nonconventional outflow pathway utilizes the uveal meshwork instead of the trabecular meshwork, with AH draining into the ciliary muscle interstitium. *Via* the

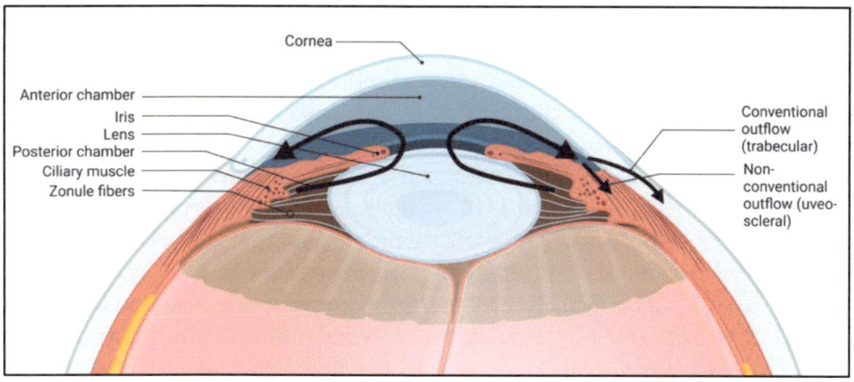

Figure 1.
Aqueous humor flow and outflow, depicting conventional outflow pathway via the trabecular meshwork that drains into the episcleral veins and nonconventional outflow via the uveoscleral pathway that drains into the ciliary interstitium.

uveal meshwork, AH enters the connective tissue between the ciliary muscle bundles and exits *via* one of three pathways, the uveoscleral, uveovortex, or uveolymphatic routes, aptly named for the vascular endpoints of the orbital vasculature, vortex veins, and ciliary lymphatics, respectively. Nonconventional outflow is largely dependent on ciliary muscle tone, as shown by experiments that use the muscarinic agonist pilocarpine to decrease nonconventional outflow. However, the definition of the nonconventional pathway has expanded to include pathways where AH leaves the eye other than through the trabecular meshwork, that is, retinal *via* the retinal pigment epithelium (typically negligible) [4].

The relative contribution of the two principal outflow pathways is difficult to determine, but studies [5–7] suggest that the conventional pathway accounts for 70–90% of total outflow. However, a key difference is that conventional outflow *via* the trabecular meshwork is pressure-dependent, while nonconventional outflow is not [8]. During times of inflammation, nonconventional pathway outflow can increase to account for up to 60% of total drainage [5]. With increasing age, the outflow utility of both pathways gradually declines, though relatively greater decline is seen in nonconventional outflow [7, 9].

The different pathways can be selectively targeted by medications and surgery to reduce production and increase outflow of aqueous in the management of glaucoma.

3. Key clinical trials

Key clinical trials have informed management of glaucoma and the utility of IOP measurement in clinical practice. In 1999, results from the Early Manifest Glaucoma Trial (EMGT) evaluated whether immediate IOP reduction in early, previously untreated open-angle glaucoma affected disease progression [10]. Participants were randomized to combined medical and laser therapy or no initial treatment, with follow-up for a median of 6 years. The treatment group saw an average IOP reduction of 5.1 mmHg (25%), with a decreased frequency of disease progression (45% versus 62%; p = 0.007) that occurred much later. Treatment was found to have a protective effect in all patients, including those with high and low IOP, young and old age, and early and late disease stage. Analysis showed that a 1 mmHg reduction in IOP from baseline was associated with a 10% reduced risk of disease progression, with later analysis suggesting central corneal thickness as a risk factor in POAG and low blood pressure as a risk factor in normal tension glaucoma (IOP less than 21 mmHg). EMGT was the first large randomized clinical trial to demonstrate the utility of immediate IOP reduction in preventing glaucomatous progression.

Prior to data from the Collaborative Normal Tension Glaucoma Study (CNTGS), it was unclear whether IOP contributed to glaucomatous optic nerve damage and visual field loss in patients whose IOPs were within the normal range. This study [11, 12], however, showed that reductions in IOP by 30% slowed the rate of 5-year glaucomatous damage in this population, similar to primary open-angle glaucoma (POAG). Higher benefit was observed in females, those with a family history of glaucoma, with mild disc excavation, without a personal history of cardiovascular disease, and without family history of stroke. Key limitations, however, were that the definition of normal tension glaucoma in this study (24 mmHg) was higher than typically defined in clinic and that central cornea thickness was not measured.

The Ocular Hypertension Treatment Study (OHTS) [13–15] sought to assess the efficacy of IOP-reduction topical pharmacotherapy in preventing initial POAG onset. Adults with an elevated IOP (24–32 mmHg) without glaucomatous damage were randomly assigned to either the observation group or the topical ocular hypotensive group, which had a goal of 20% or more IOP reduction. Over the 5-year study period, the observation group had an IOP reduction of 4.0% ± 11.6%, while the pharmacotherapy group had an IOP reduction of 22.5% ± 9.9%. At 5 years, the observation group had a 9.5% probability of progression to POAG, compared to 4.4% in the pharmacotherapy group (hazard ratio, 0.40; $P < .0001$). Thus, this study established that topical pharmacotherapy in those with IOP greater than 24 mmHg reduced the risk of POAG development by 60%. Besides IOP, other factors predicting POAG development in the trial were older age, African American race, male sex, larger vertical and horizontal cup–disc ratio, greater Humphrey visual field pattern standard deviation, heart disease, and thin central corneal thickness (thickness was 553.1 ± 38.8 μm in patients who developed POAG, versus 574.3 ± 37.8 μm in those who did not).

The Advanced Glaucoma Intervention Study (AGIS) [16, 17] reinforced the importance of IOP control in preventing glaucoma progression and visual field deterioration in eyes that had exhausted maximally tolerated pharmacotherapy and underwent surgery. After argon laser trabeculoplasty or trabeculectomy, eyes with IOP greater than 17.5 mmHg (based on three 6-month follow-up visits) had a greater visual field defect (scored from 0 to 20) at follow-up than those with an IOP less than 14.5 ($p = 0.002$), with a greater degree of defect at 7 years (1.89 units; $P < .001$) than at 2 years (0.64 units; $P = .071$). A separate analysis in the same trial showed that eyes with less than 50% of IOP readings less than 18 mmHg (based on 6-month follow-up visits over 6 years post-surgery) had a greater degree of visual field deterioration compared to those with 100% of IOP readings less than 18 mmHg ($P = .083$), with the level of deterioration worse at 7 years (1.93 units; $P < .001$) than at 2 years (0.25 units; $P = .572$). Together, these findings highlighted the role of IOP management in preventing visual field worsening for glaucoma patients, especially those with a more progressed disease state.

While these trials only offer a glimpse into the history and development of glaucoma, they demonstrate the importance of IOP in clinical practice. Whether monitoring for glaucoma onset or effectiveness of intervention, measurement of IOP offers clinicians a critical data point that holds significant predictive value.

4. Measuring IOP

There are many considerations when designing devices to measure IOP. Accuracy and reliability compared to true IOP, which is only measurable by invasive manometry, are understandably two of the most important aspects of a device design. However, ease of use for the practitioner as well as tolerability for the patient are crucial factors for devices to be widely accepted for use in the clinical setting. This section will explore the many forms of tonometry invented over the years and discuss the advantages and disadvantages of each modality.

4.1 Applanation tonometry

Applanation tonometry is widely considered the most accurate and reliable form of IOP quantification. These instruments work by controlling the amount of force required to flatten a discrete area of cornea, which is used to calculate a corresponding eye pressure using the Imbert–Fick law. This law states that the pressure inside an ideal, dry, thin-walled sphere is equal to the force needed to flatten its surface divided by the area of flattening, or $P = F/A$, where P = pressure, F = force, and A = surface area of the flattened cornea [18].

4.1.1 Goldmann applanation tonometer

First invented in 1948 by Hans Goldmann [19], the Goldmann applanation tonometer (GAT) (**Figure 2**) is still currently considered the gold standard of tonometry to this day. As a truncated cone positioned on a slit lamp, the GAT makes

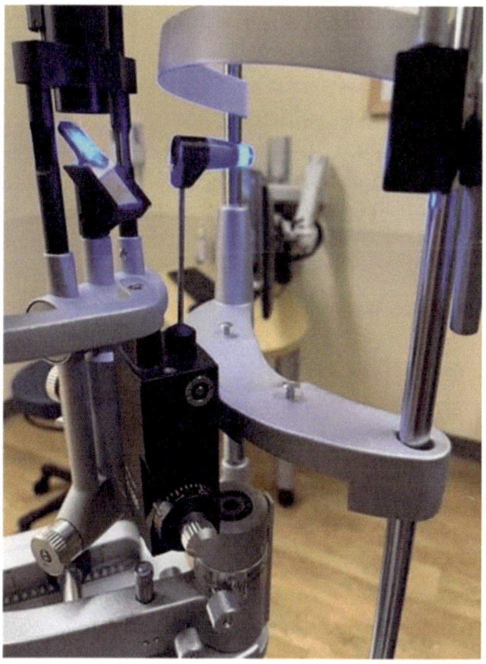

Figure 2.
The Goldmann applanation tonometer on the slit lamp.

contact with the cornea with a flattened surface area of about 7.35 mm^2. This is the area at which the tear film meniscus for the tonometer head counterbalances the resistance of the cornea to flattening. The flattening force (in grams) multiplied by 10 is what calculates the IOP (in mmHg). When using GAT in the clinic, the cornea must first be anesthetized for patient comfort. Fluorescein dye is then applied to the patient's eye so that the tear film is highlighted when blue light is shined. A dual prism embedded in the cone is used to divide the image of the tear meniscus into a superior and an inferior arc. After using peripheral knobs to adjust the force so that the two arcs align under slit lamp, the intraocular pressure can be read in mmHg. Although a very precise device, the GAT must be checked for calibration intermittently by checking whether the feeler arm is balanced at given dial positions to ensure accurate pressure readings [20]. While it is the most commonly used device due to its relative ease of use, accuracy (when calibrated properly), reproducibility, and affordability, there are a few disadvantages of this method. Various corneal parameters may affect measurement accuracy, including central corneal thickness (CCT), corneal curvature, axial length, hysteresis, and so forth. CCT most greatly impacts measurements, as the device was originally designed using the estimated average corneal thickness of 520–540 μm [21]. A study measured IOP with GAT in the central and temporal regions of the cornea before and after photorefractive keratectomy (PRK). It was found that after the central areas of the cornea were thinned by PRK, they measured an IOP 2–3 mmHg lower than the temporal regions [22]. Such studies demonstrate that thin corneas can lead to artificially low IOP measurements, whereas thick corneas may overestimate IOP by GAT. This is significant as thin corneas are a risk factor for glaucoma [23, 24]. When CCT is greater than 600 microns, a steep corneal curvature begins to have a significant impact on GAT measurements due to hysteresis and distribution of the tear film [25]. For corneas with vastly different radii between flat and steep meridians, GAT readings are recommended to be taken at both the steepest and flattest corneal axes or vertical and horizontal axes if unknown. The two readings are then averaged for the final IOP measurement [26]. Other possible limitations and errors to GAT include the tear film getting too little or too much fluorescein, having an irregular or scarred cornea, patient positioning with the slit lamp, GAT disinfection and calibration between patients, and physician experience [27].

There are also portable versions of GAT, such as Perkins and Draeger tonometers, which forgo the need for a slit lamp, allowing for supine IOP measurements for bed-bound patients.

4.1.2 Applanation resonance tonometry

Applanation resonance tonometry (ART) (**Figure 3**) also uses the principles of applanation but takes continuous measurements of force and contact area using a piezoelectric sensor. Similar to GAT, it is mounted on a slit lamp and requires anesthetic drops for the patient prior to use. The sensor contains a piezoelectric element that vibrates at a certain resonance frequency. When the sensor makes contact with the cornea using a constant force, a shift in frequency is generated, which is proportional to the contact area. This shift can then be used to calculate IOP using the Imbert–Fick law given the constant force and measured contact area [28, 29]. Multiple points are measured, and the median measurement and a quality index reflecting standard deviation are given. This feature of the device theoretically makes ART more accurate and precise than GAT. However, past studies [30, 31] seem to indicate that ART overestimates IOP compared to GAT. This device has many of the

Figure 3.
The bioresonator Applanation resonance tonometer. Photo from Brusini and colleagues [27].

same disadvantages as GAT in practice, including measurements being influenced by CCT and corneal biomechanics, use of the slit lamp and anesthetics, and the need to sanitize the probe between patients.

4.1.3 Noncontact tonometry

Noncontact tonometry (NCT) or air-puff tonometry (**Figure 4**) uses the principles of applanation but does not require anesthesia or fluorescein drops. NCT works by gradually pulsing air at an increasingly strong force until the cornea is flattened [32]. At that point, the production of air is stopped, and the force required to flatten is recorded and used to calculate IOP. The disadvantages of the NCT are that it becomes less accurate at higher subject IOPs, especially with IOP > 20 mmHg [33]. It is also influenced by corneal properties, similar to GAT [34]. Results also seem increasingly variable depending on the device brand and model and require regular calibration [35]. Airborne infections could also be greater with this mode of tonometry, although it limits risk of infection in the forms of contaminated drops or device surfaces [36]. The advantages of NCT lie in its ease of use and portability of many devices. It can be more easily used by medical staff without slit-lamp experience and more tolerable for patients who are less compliant with slit-lamp positioning, such as children or patients with disabilities. This modality could be more useful as a screening tool for patients without suspicion or risk factors of increased IOP or glaucoma undergoing routine

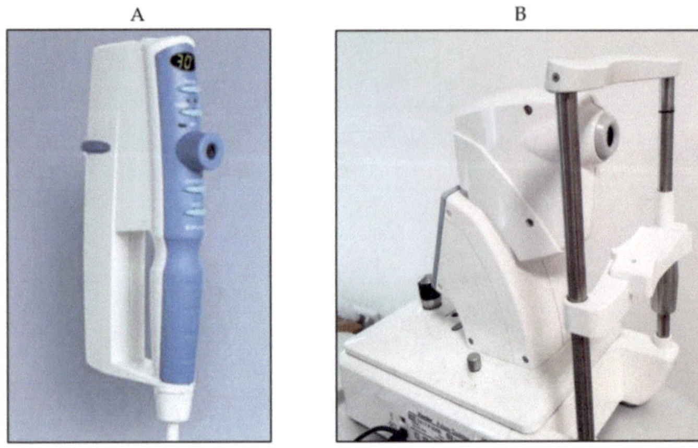

Figure 4.
The Pulsair EasyEye and Pulsair desktop tonometer. Photo from Brusini and colleagues [27].

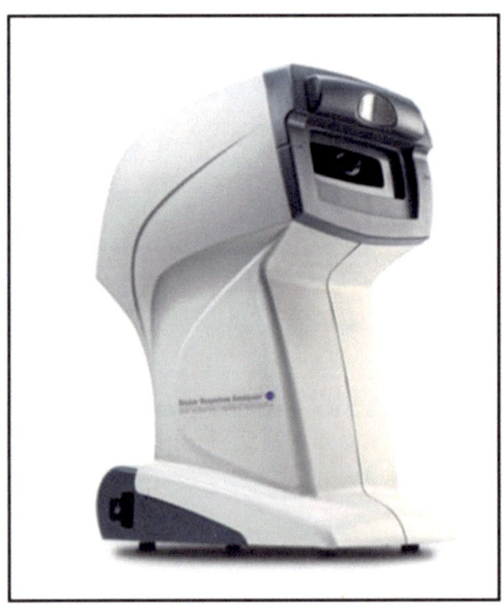

Figure 5.
The ocular response Analyzer G3. Photo from Brusini and colleagues [27].

checkups. It can be a useful alternative when GAT measurements may be difficult or skewed, such as for patients with limited cooperation, ocular pain, increased tear film meniscus size, or postoperative patients with lid edema.

4.1.4 Ocular response Analyzer

The Ocular Response Analyzer (ORA) (**Figure 5**) is a new version of the NCT. It utilizes an optical electrical system to additionally measure corneal elasticity or hysteresis to calculate a "corrected" IOP less dependent on CCT and other corneal parameters. The corneal elasticity is measured by taking the difference between the initial applanation force measured and a second applanation force [37]. The second applanation point is measured after fully indenting the cornea with stronger air columns and then slowly decreasing the air pressure to allow the cornea to "reinflate" until it reaches the second applanation point [37]. The pros of the ORA are its ease of use of noncontact nature and it had also been shown to be less influenced by corneal biomechanics and more accurately measures IOP after refractive surgery of the cornea when compared to GAT [38]. It has also been shown to better predict rates of glaucoma progression. This is due to the fact that corneal hysteresis is related to various glaucoma factors, including high cup-to-disc ratio and visual field defects [39, 40]. It can even help detect patients with corneal pathologies, such as keratoconus, or those at risk for corneal ectasia after refractive LASIK surgery [41, 42]. In terms of cons, the ORA seems to overestimate IOP, particularly at high IOP values [43, 44]. It is also a rather expensive device.

4.2 Indentation tonometry

Indentation tonometry is based on the principle that a force will sink into a soft eye further than into a hard eye.

4.2.1 Schiøtz tonometer

The first instrument that applied this principle was the Schiøtz tonometer (**Figure 6**), which is no longer in use in the modern-day clinical setting. To use this instrument, the patient had to lie in a supine position and have their cornea indented by a plunger loaded with weights ranging from 5 to 15 grams. The depth of indentation into the cornea, ranging from 0 to 10 mm, is indirectly proportional to and converted to IOP. Additionally, the coefficient of ocular rigidity, unique to an individual's eye, had to be accounted for to measure a more accurate IOP. The rudimentary measurement method using weights makes accurate IOP measurements difficult. It is cumbersome to use, and improper positioning of the eye, variability of ocular rigidity, and instrument variability all make the precision and reliability of this device questionable. Furthermore, this instrument is not practical without anesthesia due to patient tolerability. Although it is relatively affordable, simple, and does not require electronics, the development of more accurate, precise, and tolerable instruments has made the Schiøtz tonometer obsolete in modern-day clinics and only used in remote, low-resource settings [45].

4.2.2 Corvis ST

The Corvis ST (**Figure 7**) is a more novel noncontact device that uses indentation tonometry principles using a jet of air. It has a Scheimpflug camera that monitors an 8.5 mm diameter at the center of the corneal surface. It visualizes with a high resolution of more than 4300 frames per second the corneal deformation and its return to normal shape as the air-jet indents the cornea. It then characterizes corneal deformation parameters to produce a "biomechanically corrected" IOP [46]. As a result, it has shown to be less impacted by CCT and corneal properties and more accurate when measuring patients who had previously undergone refractive surgery [47]. It has also proven to be a precise method of measuring IOP in healthy subjects [48]. It has similar advantages as NCT and ORA due to its noncontact nature, and it further

Figure 6.
The Schiøtz tonometer. Photo from Brusini and colleagues [27].

characterizes corneal parameters. Its disadvantages include its need for specialized training for use, table mount setup, and high cost.

4.3 Combined applanation and indentation tonometry

4.3.1 Pneumotonometer

The pneumotonometer (**Figure 8**) combines indentation tonometry with applanation principles. To applanate the cornea, a 5 mm diameter silicone tip indents the cornea using pressure of a controlled flow of air. IOP is measured at the equilibrium point at which both the tip and cornea are flat [49]. Pneumotonometry has been shown to be quite accurate and precise for glaucoma screening with greater reliability than even GAT after corneal procedures, such as PRK and LASIK [50–52]. Its pros include its portability, minimal contact with the cornea, and ease of use. Unfortunately, the silicone tip can be difficult to disinfect, and it requires anesthetics for use and calibration for reliable readings. The pneumotonometer was also found to underestimate IOP at lower values and overestimate IOP at higher values compared to GAT and can also be easily influenced by CCT [53, 54].

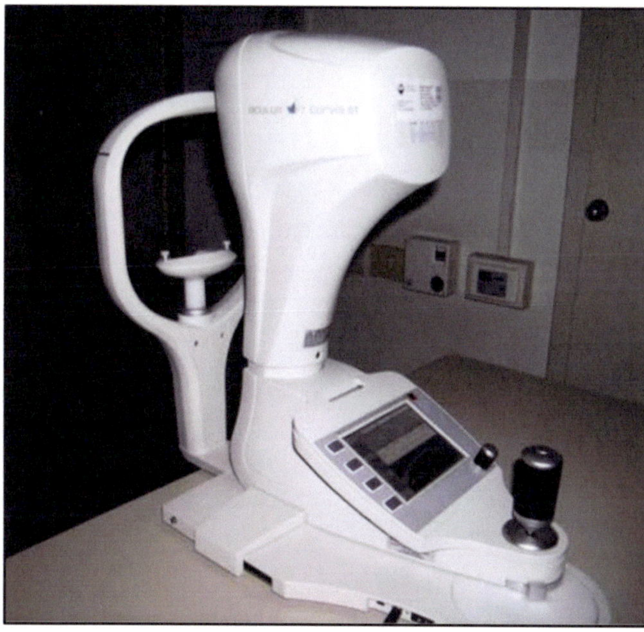

Figure 7.
The Corvis ST tonometer. Photo from Brusini and colleagues [27].

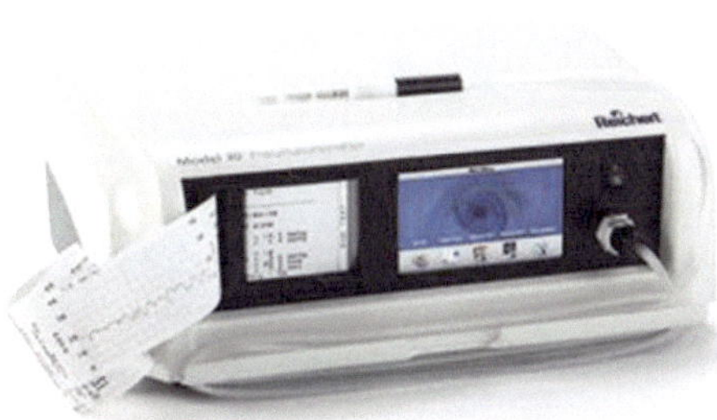

Figure 8.
The Reichert Pneumotonometer. Photo from Brusini and colleagues [27].

4.3.2 Tono-pen

Another device that applies both applanation and indentation tonometry is the Tono-Pen (**Figure 9**). The Tono-Pen is a small portable battery-powered device that uses a footplate with a tiny plunger connected to a strain gauge on its applanation surface. As the plunger makes contact with the cornea, it experiences increasing resistance due to the IOP, which is recorded as force on the strain gauge. When applanation occurs, the steadily increasing force decreases momentarily because the force is shared by the footplate and the plunger. This force value is then recorded and used to calculate IOP with the known applanation area. Multiple readings are recorded and averaged to produce the final reading in mmHg with a standard deviation value displayed as its reliability. This device is popular in the modern setting because it is handheld, can be used in any position, and does not require special training [49]. Because of its small area of contact, it can reliably measure irregular corneas and eyes

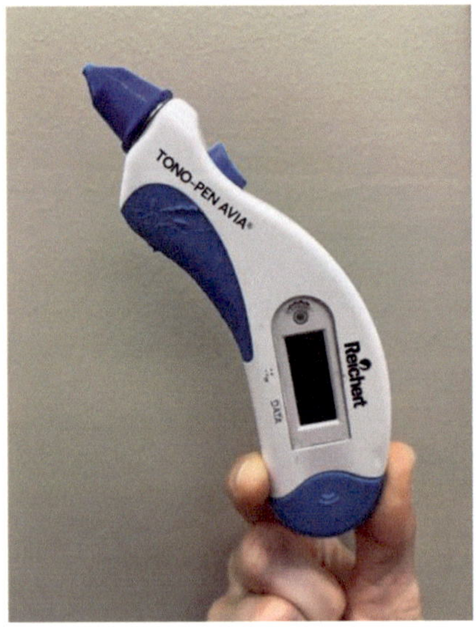

Figure 9.
The Tono-pen AVIA handheld tonometer.

with therapeutic contact lenses without removing the lenses [55, 56]. It also uses a disposable latex cap to reduce risk of infection between patients. Despite its convenience, the Tono-Pen is significantly impacted by CCT and seems to underestimate or overestimate IOP [31, 54, 57].

4.4 Rebound tonometry

Rebound tonometry measures IOP as a function of a probe's deceleration as it contacts the cornea.

4.4.1 iCare

The main rebound tonometer in use is the iCare tonometer (**Figure 10**). It uses a magnetized probe that is propelled toward the cornea and decelerates as it makes contact. The probe decelerates more quickly if the IOP is high and more slowly if the IOP is low. The motion of the probe generates a voltage in an internal solenoid that is then used to calculate IOP by a microprocessor [27]. The final IOP is averaged from several consecutive measurements. iCare has shown to have strong concordance with GAT of within 2–3 mmHg for both normal and glaucoma patients [58, 59]. Some studies do show, however, that iCare may underestimate pressures for IOPs greater than 23 mmHg [60]. It is also influenced by corneal parameters, particularly estimating higher IOP with thicker corneas [61–64]. Positioning of the tip may also influence measurements. It seems to be most accurate for mid-range levels of IOP, ranging from 16 to 23 mmHg [62]. This device has also been demonstrated to make reliable and repeatable measurements with an intraclass correlation of >0.9 [65, 66]. Overall, the iCare tonometer is quick, affordable, easy to use, portable, and can be used in any position. It does not require anesthetics and is

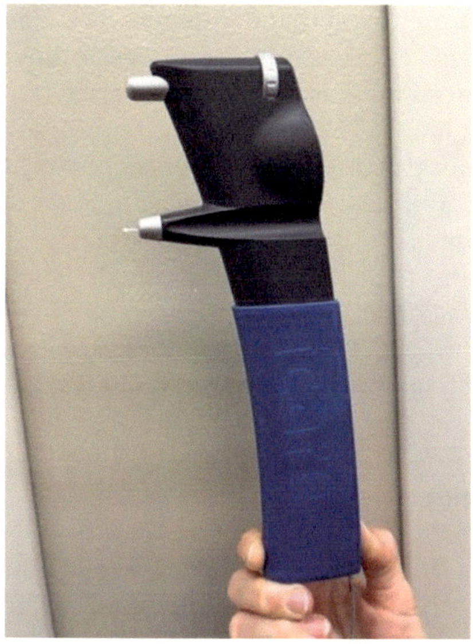

Figure 10.
iCare rebound tonometer, with attached disposable probe.

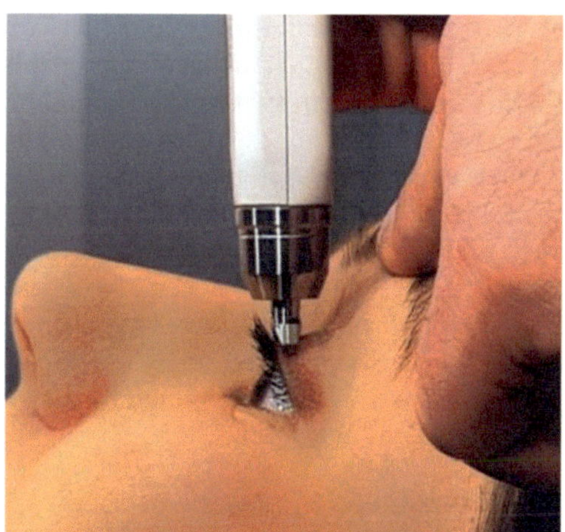

Figure 11.
Diaton tonometer for transpalpebral sceral palpation. Figure from Berg and colleagues [67].

well tolerated by patients. It has minimal contact and disposable tips, reducing the risk of infection or corneal damage and allowing for postoperative measurements. Newer versions have even been marketed with the intention of allowing self IOP measurements at home.

4.4.2 Transpalpebral scleral palpation

Transpalpebral scleral palpation (**Figure 11**) applies the principles of rebound tonometry through the upper eyelid to estimate IOP. The Diaton tonometer is one such device that releases a metal rod that decelerates as it contacts the eyelid, superior tarsal plate, and superior sclera. Its advantages are its noninvasive nature, reduced risk of infection, lack of need for anesthesia, and portability, which may make IOP measurements possible at home. The downside is its accuracy, only seeming to be a better alternative to finger palpation and tactile tonometry, which are subjective and qualitative measures of pressure [68]. Transpalpebral scleral palpation seems to better estimate IOP in thinned cornea after photorefractive surgeries compared to GAT [69]. However, it seems to be less sensitive overall than GAT in measuring IOP for patients with glaucoma [70, 71]. It overestimates IOP in the lower IOP ranges and underestimates in higher ranges. There is also poor precision and increased variability between readings for this modality, which make it questionable for legitimate clinical evaluation of IOP [68].

4.5 Dynamic contour tonometry

The Dynamic Contour Tonometer (DCT) (**Figure 12**) implements the Pascal principle to calculate IOP [72]. According to the Pascal principle, in an enclosed space, changes in pressure are applied to all parts of a fluid. The device is mounted on a slit-lamp and measures pressure directly with a small piezoelectric sensor [73] as it touches the corneal surface, theoretically uninfluenced by corneal properties [74–77]. As a result, it can be used for patients who have had photorefractive surgeries. It can also measure ocular pulse amplitude, which can indirectly characterize choroidal perfusion, an important factor in the onset and progression of glaucoma [78]. The DCT has even been shown to have higher reproducibility and precision compared to GAT. It generates a quality score from 1 (optimal) to 5 (unacceptable) to indicate the quality of IOP measured with a score of 1 or 2 being reliable for clinical practice [79, 80]. While largely concordant, DCT measurements do seem to measure much higher than GAT, particularly at low CCT [81]. The main disadvantages of the DCT include the training necessary for use and complexity for the patient, as cooperation with optimal head and eye positioning for a minimum of 8 seconds is needed. The device also requires anesthetic drops, corneal contact, and a slit lamp. Fortunately, the sensor tip is protected by disposable sensor caps to reduce the risk of infection.

A B

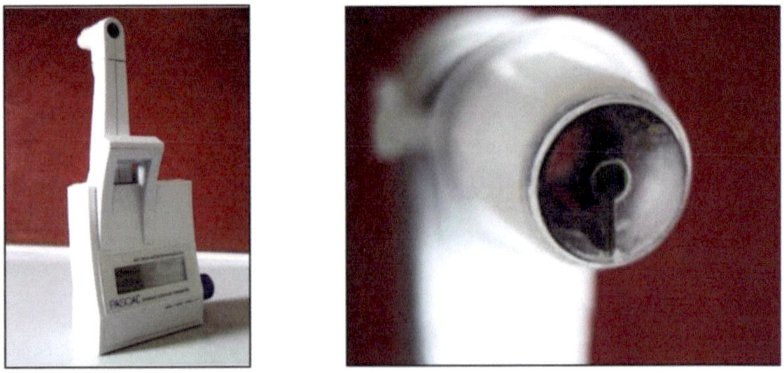

Figure 12.
The dynamic contour tonometer. Photo from Brusini and colleagues [27].

5. Measuring IOP: Looking ahead

While advances in technology have supported innovation in implantation and indentation tonometry, a key shortcoming is that these methods are instantaneous—only IOP at a singular point in time is measured. It is well noted, however, that IOP can fluctuate. This fluctuation can be 4–5 mmHg in healthy individuals and even higher in glaucomatous eyes. The role of long-term IOP variation as a glaucoma risk factor has been well established in clinical trials, as previously discussed. Several studies [82–85] suggest that diurnal-nocturnal IOP fluctuations may be also indicative of glaucoma progression. In assessing the role of 24-h IOP changes using a contact lens sensor on the rate of visual field progression in glaucomatous eyes, a 2016 investigation [85] found that IOP parameters—which included the number of large IOP peaks and the mean peak ratio, which accounts for the magnitude of the peaks and time to peak—could predict the rate of glaucoma progression. Static IOP measurements do not enable measurement of these parameters and may misdiagnose patients. This can be particularly problematic for normo-tensive glaucoma patients, who demonstrate significant damage and progression despite normal IOP values in clinic. For this group, an elevated IOP may be found during non-clinic hours, such as at night or early in the morning. Although clinical methods gain partial insight into diurnal IOP variations by obtaining static IOP measurement in the morning and at night, the appreciation of IOP as a dynamic measure has enhanced interest in continuous tonometry to monitor disease. Several devices have been developed for continuous IOP monitoring, ranging from contact lens sensors to surgical implants. With only a few studies to assess the accuracy and precision of these monitors, a thorough evaluation of evidence is needed to ascertain indications for and limitations of use.

5.1 Contact lens sensors: Sensimed triggerfish

The Sensimed Triggerfish (**Figure 13**) is a CE-marked and FDA-approved contact lens sensor (CLS) designed for 24-h IOP monitoring. At 14.1 mm in diameter and 585 um in central thickness, the Sensimed Triggerfish silicone contact lens contains a

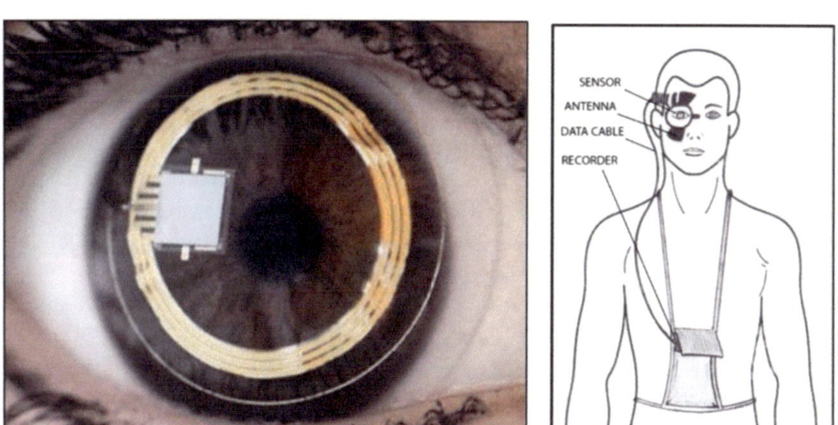

Figure 13.
(left) – Sensimed Triggerfish contact lens sensor worn in sample eye; (right) – The contact lens sensor wirelessly transmits data to an adhesive periorbital patch that includes the antenna. Via a data cable, the periorbital patch can then transmit data to a handheld recorder device. Figure from Brusini et al. [27].

microprocessor, two strain gauges, and an antenna. The strain gauges detect changes in corneal shape, for which a correlation between corneal curvature and IOP has been demonstrated in animal models [86, 87]. Strain gauge readings are taken for a 30-second period every 5 minutes for the duration of the 24 h. The strain gauge transmits data to a wireless periorbital antenna attached to the patient, which then further transmits data to a portable recorder. The IOP data is recorded in millivolts and is measured relative to the first measurement, which is taken as zero.

Studies suggest that the Triggerfish device provides reproducible data. Mansouri and colleagues [88] spaced 24-h IOP readings one week apart in 40 patients, with a Pearson correlation coefficient of 0.59 between sessions, interpreted by the authors as fair reproducibility. Hollo and colleagues [89] reported a Pearson coefficient of 0.729 between CLS curves in their study of 9 patients. While data of reproducibility is promising, a major limitation of this device is that the correlation between device output in millivolts and IOP, in mmHg, is not established. No algorithm has been able to convert CLS output directly into IOP, even in animal models, as different rates of CLS output were noted per IOP change in different eyes. In enucleated pig eyes, Leonardi and colleagues [86] noted changes in CLS output ranging from 0.067 mV to 0.124 mV per mm Hg change in IOP. It is even more difficult to assess validity in human eyes, as insertion of the CLS precludes traditional tonometry. Researchers have attempted to circumvent this by utilizing the CLS in one eye and measuring IOP in the other eye. In one study [87] of 33 patients utilizing the CLS in one eye and IOP measured every 2 h *via* pneumotonometry in the other eye, the Pearson correlation coefficient was 0.956. Despite their promising data on CLS reproducibility, Hollo and colleagues [89] did not find a correlation between change in CLS output over 24 h and change in IOP as measured by traditional GAT before and after CLS use ($r = -0.223$). These studies highlight the questions surrounding the validity of the Sensimed Triggerfish CLS.

The principal advantage of this device is its noninvasive ease of use, as many patients are familiar with traditional contact lenses. Accordingly, it is generally well-tolerated by both glaucomatous and non-glaucomatous patients. Typically, two short clinic visits are required for Triggerfish use, one visit to apply the device and one to remove the device. Studies using the visual analog scale (VAS), where 0 represents no discomfort and 100 represents severe discomfort, suggest high levels of patient tolerability. For example, a study [90] of 20 glaucoma and 20 non-glaucoma patients showed a VAS score of 21.82 and 26.8, respectively. Clinical studies [88, 90] show 75–95% of patients had a device-related adverse event, though the majority of complications were mild. Complications associated with Sensimed Triggerfish use were hyperemia, blurred vision, and punctate keratitis [89, 91]. In one study [88], only 3% of patients had severe complications, all of which were conjunctival hyperemia. All complications resolved within 24 h of device removal. Moreover, artificial tears may be utilized in the event of mild ocular pruritus.

5.2 Implantable microsensors: Eyemate-IO and eyemate-SC

While 24-h IOP monitoring *via* CLS is promising, patients may benefit from continuous IOP monitoring for longer periods of time. Implantable microsensors, such as the Eyemate-IO (**Figure 14**), have been developed that enable permanent continuous monitoring of IOP. Developed by Implandata Ophthalmic Products, the nonmagnetic Eyemate-IO device is surgically inserted into the ciliary sulcus posterior to the iris and anterior to an artificial lens in cataract surgery. Eight pressure- and temperature-sensitive capacitors are attached to a circular golden antenna. Each

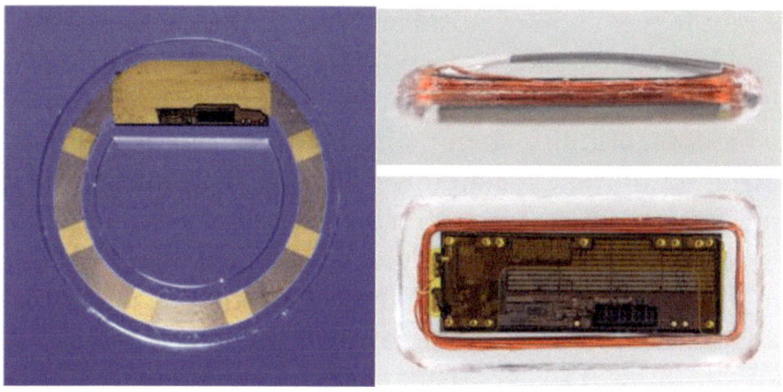

Figure 14.
(left) – EYEMATE-IO, for implantation into the ciliary sulcus; (right, top and bottom) – EYEMATE-SC, for implantation into the suprachoroidal space. Photos from Choritz et al. [92] and Szurman et al. [93].

capacitor contains two parallel plates, wherein the distance between the two plates reflects a corresponding IOP [94]. The device weighs 0.1 g, with an outer diameter of 1.3 mm, inner diameter of 7 mm, and thickness of 0.9 mm. The readings captured by the transducer are transferred to a handheld device, which also powers the intraocular device *via* electromagnetic coupling. A smartphone app can be used to review IOP history and set medication alerts. Through the Eyemate-IO, up to 10 readings per second can be taken and the device can measure IOP on demand by bringing the external device close to the eye. As this device is placed inside the eye, factors that influence the accuracy of applanation tonometry, such as corneal rigidity and central corneal thickness, do not play a role.

The ARGOS-01 study evaluated the Eyemate-IO device in six patients with POAG or normotensive glaucoma receiving cataract surgery. Upon implantation, four of the patients developed perioperative inflammation that was well controlled with steroids. One-year results [95] showed that all participants maintained glaucoma control. Mild pupillary distortion that remained stable over time was reported in all patients, but no other adverse events were noted. Furthermore, central corneal thickness and endothelial cell count were stable in all six patients. However, results of the ARGOS-01 trial showed that IOP as measured by the implanted sensor may drift higher.

This prompted the creation of a second-generation Eyemate device, whose 12-month outcomes were evaluated in the ARGOS-02 trial [92]. This trial of 22 patients demonstrated minimal surgical complication (most commonly, a floppy iris) with an incision size of 5.5 to 6 mm required. Compared to the first-generation design, decreases in pigment dispersion and transillumination defects were seen. Like the first iteration, all implantation-related adverse events resolved promptly with appropriate treatment. 12 months of follow-up showed correlation between device readings and GAT results, with an interclass correlation coefficient of 0.783. However, Eyemate-IO IOP outputs were on average 3.2 mmHg higher than GAT readings.

While the second-iteration Eyemate-IO had several improvements, key limitations still existed due to device placement in the ciliary sulcus: iris chafing, iris atrophy, pupillary distortion, and pigment dispersion [92]. Moreover, implantation was limited to pseudophakic patients, thus excluding many patients. Accordingly, an IOP monitor for suprachoroidal implantation, the Eyemate-SC (**Figure 14**), was developed with the major advantage that the anterior chamber remains intact with readings still independent of corneal mechanics. More compact than the Eyemate-IO, the Eyemate-SC showed promising results in animal models [96]. Implantation of the device in six rabbit eyes showed no adverse outcomes, with histological analysis demonstrating mild fibrosis without any pathology. Importantly, there was strong correlation with device IOP and intracameral IOP up to 30 weeks following implantation with strong biocompatibility. In humans, Szurman and colleagues [97] evaluated safety and performance of the Eyemate-SC in 23 eyes that underwent canaloplasty or deep sclerectomy. Implantation was successful in all eyes without major complication, choroid injury, or bleeding. Touch-sensitivity was the only device-related adverse event during the first six months, reported in three patients. Temporary increases in corneal astigmatism were reported in the early postoperative period. While device IOP and GAT IOP did show transient deviations in agreement in the first few weeks after implantation, these discrepancies normalized after three months with an agreement of −0.15 mm Hg ± 2.28 between GAT IOP and Eyemate-SC IOP after six months. It is likely that this increased initial discrepancy is due to corneal astigmatism caused by device implantation of the device, which has been shown to affect GAT IOP values [98]. 2023 results [93] of the 12-month follow-up period of the same cohort continued to show promise, with no evidence of device migration or major adverse events from the device. In terms of Eyemate-SC accuracy, mean IOP difference between GAT and Eyemate-SC was 0.8 mmHg, with 8.3% of measurements showing a discrepancy of >5 mmHg. As the eye healed following surgery, readings improved in accuracy. From three months to twelve months, the mean difference between GAT and device telemetry readings was −0.2 mmHg, with all device measurements within 5 mmHg of GAT. Overall, one-year outcomes of the Eyemate-SC point toward promising outcomes in safety, tolerability, and efficacy.

5.3 Injectable microsensor: IOP connect

While the Eyemate devices require surgical implantation, IOP Connect by Injectsense is developed for outpatient implantation *via* intravitreal injection or as an intraocular lens-embedded device [99]. At 4.6 mm x 1 4 mm x 0.6 mm, the implantable iteration of the device consists of a pressure sensor microelectromechanical systems (MEMS) chip, an ASIC chip that is connected to the pressure sensor for telemetry and charging, and a solid state microbattery chip. The device captures IOP readings at intervals specified by the physician. Once a week, the patient must utilize a wearable receiver, in the form of smart glasses, for a few minutes to enable data upload and charging. IOP Connect received FDA Breakthrough Device Program designation in 2020. No results have been published regarding device efficacy to date.

6. Conclusion

The importance of IOP in ophthalmologic practice cannot be understated. Investigations into the physiology of aqueous humor dynamics and several clinical trials have identified IOP measurement and monitoring as the mainstay for knowing when to intervene for numerous pathologies, such as glaucoma. Appreciation of applanation principles enabled development of the GAT, which became the gold standard for IOP measurement. Over time, numerous instruments have been developed using the principles of applanation and indentation. As data-driven medicine evolves, continuous IOP monitoring will better illuminate variations in IOP over time. Several continuous IOP sensors have been developed, including contact lens sensors for 24-h

DOI: http://dx.doi.org/10.5772/intechopen.1003876

use and suprachoroidal implants for long-term monitoring. While significant clinical trials are still needed to validate continuous IOP monitors, they represent a promising therapeutic tool that may revolutionize the management of glaucoma. Of the many devices that have been developed, each has their distinct strengths and weaknesses in terms of accuracy, reliability, practicality, and cost. Ultimately, it is up to the clinician to choose the device that best suits the unique circumstances of their practice and their patients.

Conflict of interest

The authors declare no conflict of interest.

Author details

Sohum Sheth[1], Kevin Peng[1], Ankit Shah[2,3] and Mark Disclafani[2]*

1 College of Medicine, University of Florida, Gainesville, FL, USA

2 Department of Ophthalmology, College of Medicine, University of Florida, Gainesville, FL, USA

3 Department of Veterans Affairs, Division of Ophthalmology, Salem, VA, USA

*Address all correspondence to: mdisclafani1@gmail.com

References

[1] Goel M, Picciani RG, Lee RK, Bhattacharya SK. Aqueous humor dynamics: A review. Open Ophthalmology Journal. 2010;**47**:52-59

[2] Maus TL, Brubaker RF. Measurement of aqueous humor flow by fluorophotometry in the presence of a dilated pupil. Investigative Ophthalmology & Visual Science. 1999;**40**(2):542-546

[3] Brubaker RF. Chapter 9 clinical measurements of aqueous dynamics: Lmplications for addressing Glaucoma. In: Current Topics in Membranes. Elsevier; 1997. pp. 233-284. Available from: https://linkinghub.elsevier.com/retrieve/pii/S007021610860249X

[4] Sunderland DK, Sapra A. Physiology, Aqueous Humor Circulation. Treasure Island (FL): StatPearls Publishing; 2023. Available from: http://www.ncbi.nlm.nih.gov/books/NBK553209/

[5] Fautsch MP, Johnson DH. Aqueous humor outflow: What do we know? Where will it lead us? Investigative Ophthalmology & Visual Science. 2006;**47**(10):4181-4187

[6] Weinreb RN. Uveoscleral outflow: The other outflow pathway. Journal of Glaucoma. 2000;**9**(5):343-345

[7] Alm A, Nilsson SFE. Uveoscleral outflow: A review. Experimental Eye Research. 2009;**88**(4):760-768

[8] Brubaker RF. Measurement of uveoscleral outflow in humans. Journal of Glaucoma. 2001;**5**(Suppl. 1): S45-S48

[9] Johnson M, McLaren JW, Overby DR. Unconventional aqueous humor outflow: A review. Experimental Eye Research. 2017;**158**:94-111

[10] Heijl A. Reduction of intraocular pressure and glaucoma progression: Results from the early manifest glaucoma trial. Archives of Ophthalmology. 2002;**120**(10):1268

[11] Anderson DR, Normal Tension Glaucoma Study. Collaborative normal tension glaucoma study. Current Opinion in Ophthalmology. 2003;**14**(2):86-90

[12] Anderson DR, Drance SM, Schulzer M, Collaborative Normal-Tension Glaucoma Study Group. Factors that predict the benefit of lowering intraocular pressure in normal tension glaucoma. American Journal of Ophthalmology. 2003;**136**(5):820-829

[13] Kass MA, Heuer DK, Higginbotham EJ, Johnson CA, Keltner JL, Miller JP, et al. The Ocular Hypertension Treatment Study: A randomized trial determines that topical ocular hypotensive medication delays or prevents the onset of primary open-angle glaucoma. Archives of Ophthalmology. 2002;**120**(6):701-713 discussion 829-830.

[14] Higginbotham EJ, Gordon MO, Beiser JA, Drake MV, Bennett GR, Wilson MR, et al. The Ocular Hypertension Treatment Study: Topical medication delays or prevents primary open-angle glaucoma in African American individuals. Archives of Ophthalmology. 2004;**122**(6):813-820

[15] Budenz DL, Anderson DR, Feuer WJ, Beiser JA, Schiffman J, Parrish RK, et al. Detection and prognostic significance of optic disc hemorrhages during the ocular hypertension treatment Study. Ophthalmology. 2006;**113**(12):2137-2143

[16] Ederer F, Gaasterland DE, Sullivan EK, Investigators AGIS. The advanced Glaucoma intervention Study (AGIS): 1. Study design and methods and baseline characteristics of study patients. Controlled Clinical Trials. 1994;**15**(4):299-325

[17] The Advanced Glaucoma Intervention Study (AGIS): 7. The relationship between control of intraocular pressure and visual field deterioration.The AGIS Investigators. American Journal of Ophthalmology. 2000;**130**(4):429-440

[18] Allingham RR. Shields Textbook of glaucoma. Seventh ed. Philadelphia: Wolters Kluwer; 2021

[19] Goldmann H, Schmidt T. Applanation tonometry. Ophthalmologica. 1957;**134**(4):221-242

[20] Cordero I. How to verify the calibration of Goldmann tonometers. Community Eye Health. 2012;**25**(79-80):65

[21] Ehlers N, Bramsen T, Sperling S. Applanation tonometry and central corneal thickness. Acta Ophthalmologica. 1975;**53**(1):34-43

[22] Schipper I, Senn P, Niesen U. Are we measuring the right intraocular pressure after excimer laser photorefractive laser keratoplasty in myopia? Klinische Monatsblätter für Augenheilkunde. 1995;**206**(5):322-324

[23] Brandt JD, Beiser JA, Kass MA, Gordon MO. Central corneal thickness in the ocular hypertension treatment Study (OHTS). Ophthalmology. 2001;**108**(10):1779-1788

[24] Doughty MJ, Zaman ML. Human corneal thickness and its impact on intraocular pressure measures: A review and meta-analysis approach. Survey of Ophthalmology. 2000;**44**(5):367-408

[25] Matsumoto T, Makino H, Uozato H, Saishin M, Miyamoto S. The influence of corneal thickness and curvature on the difference between intraocular pressure measurements obtained with a non-contact tonometer and those with a Goldmann Applanation tonometer. Japanese Journal of Ophthalmology. 2000;**44**(6):691

[26] Mark HH, Mark TL. Corneal astigmatism in applanation tonometry. Eye (London, England). 2003;**17**(5):617-618

[27] Brusini P, Salvetat ML, Zeppieri M. How to measure intraocular pressure: An updated review of various tonometers. Journal of Clinical Medicine. 2021;**10**(17):3860

[28] Stamper RL. A history of intraocular pressure and its measurement. Optometry and Vision Science. 2011;**88**(1):E16-E28

[29] Eklund A, Hallberg P, Lindén C, Lindahl OA. An applanation resonator sensor for measuring intraocular pressure using combined continuous force and area measurement. Investigative Ophthalmology & Visual Science. 2003;**44**(7):3017-3024

[30] Ottobelli L, Fogagnolo P, Frezzotti P, De Cillà S, Vallenzasca E, Digiuni M, et al. Repeatability and reproducibility of applanation resonance tonometry: A cross-sectional study. BMC Ophthalmology. 2015;**15**:36

[31] Salvetat ML, Zeppieri M, Tosoni C, Brusini P. Repeatability and accuracy of applanation resonance tonometry in healthy subjects and patients with glaucoma. Acta Ophthalmologica. 2014;**92**(1):e66-e73

[32] Grolman B. A new tonometer system. American Journal of Optometry and Archives of American Academy of Optometry. 1972;**49**(8):646-660

[33] Stock RA, Ströher C, Sampaio RR, Mergener RA, Bonamigo EL. A comparative Study between the Goldmann Applanation tonometer and the non-contact air-puff tonometer (Huvitz HNT 7000) in Normal eyes. Clinical Ophthalmology. 2021;**15**:445-451

[34] Kyei S, Assiamah F, Kwarteng MA, Gboglu CP. The association of central corneal thickness and intraocular pressure measures by non-contact tonometry and Goldmann Applanation tonometry among glaucoma patients. Ethiopian Journal of Health Sciences. 2020;**30**(6):999-1004

[35] Atkinson PL, Wishart PK, James JN, Vernon SA, Reid F. Deterioration in the accuracy of the pulsair non-contact tonometer with use: Need for regular calibration. Eye (London, England). 1992;**6**(Pt 5):530-534

[36] Britt JM, Clifton BC, Barnebey HS, Mills RP. Microaerosol formation in noncontact "air-puff" tonometry. Archives of Ophthalmology. 1991;**109**(2):225-228

[37] Luce DA. Determining in vivo biomechanical properties of the cornea with an ocular response analyzer. Journal of Cataract and Refractive Surgery. 2005;**31**(1):156-162

[38] Zhang H, Sun Z, Li L, Sun R, Zhang H. Comparison of intraocular pressure measured by ocular response analyzer and Goldmann applanation tonometer after corneal refractive surgery: A systematic review and meta-analysis. BMC Ophthalmology. 2020;**20**(1):23

[39] Lascaratos G, Garway-Heath DF, Russell RA, Crabb DP, Zhu H, Hirn C, et al. Intraocular pressure (IOP) measured with the ocular response analyzer is a better predictor of glaucoma progression than Goldmann IOP in the United Kingdom Glaucoma treatment Study (UKGTS). Investigative Ophthalmology & Visual Science. 2014;**55**(13):128-128

[40] Susanna BN, Ogata NG, Daga FB, Susanna CN, Diniz-Filho A, Medeiros FA. Association between rates of visual field progression and intraocular pressure measurements obtained by different tonometers. Ophthalmology. 2019;**126**(1):49-54

[41] Kirgiz A, Karaman Erdur S, Atalay K, Gurez C. The role of ocular response analyzer in differentiation of Forme Fruste Keratoconus from corneal astigmatism. Eye & Contact Lens. 2019;**45**(2):83-87

[42] Randleman JB. Post-laser in-situ keratomileusis ectasia: Current understanding and future directions. Current Opinion in Ophthalmology. 2006;**17**(4):406-412

[43] Kynigopoulos M, Schlote T, Kotecha A, Tzamalis A, Pajic B, Haefliger I. Repeatability of intraocular pressure and corneal biomechanical properties measurements by the ocular response analyser. Klinische Monatsblätter für Augenheilkunde. 2008;**225**(5):357-360

[44] Martinez-de-la-Casa JM, Garcia-Feijoo J, Fernandez-Vidal A, Mendez-Hernandez C, Garcia-Sanchez J. Ocular response analyzer versus Goldmann applanation tonometry for intraocular pressure measurements. Investigative Ophthalmology & Visual Science. 2006;**47**(10):4410-4414

[45] Albert DM, Keeler R. The pressure: Before and after Schiøtz. Ophthalmology Glaucoma. 2020;**3**(6):409-413

[46] Ambrósio R Jr, Ramos I, Luz A, Faria FC, Steinmueller A, Krug M, et al. Dynamic ultra high speed Scheimpflug imaging for assessing corneal biomechanical properties. Revista Brasileira de Oftalmologia. 2013;**72**(2):99-102

[47] Bao F, Huang W, Zhu R, Lu N, Wang Y, Li H, et al. Effectiveness of the Goldmann applanation tonometer, the dynamic contour tonometer, the ocular response Analyzer and the Corvis ST in measuring intraocular pressure following FS-LASIK. Current Eye Research. 2020;**45**(2):144-152

[48] Lopes BT, Roberts CJ, Elsheikh A, Vinciguerra R, Vinciguerra P, Reisdorf S, et al. Repeatability and reproducibility of intraocular pressure and dynamic corneal response parameters assessed by the Corvis ST. Journal of Ophthalmology. 2017;**2017**:8515742

[49] Aziz K, Friedman DS. Tonometers-which one should I use? Eye (London, England). 2018;**32**(5):931-937

[50] Guildford J, O'Day DM. Applanation pneumotonometry in screening for glaucoma. Southern Medical Journal. 1985;**78**(9):1081-1083

[51] Abbasoglu OE, Bowman RW, Cavanagh HD, McCulley JP. Reliability of intraocular pressure measurements after myopic excimer photorefractive keratectomy. Ophthalmology. 1998;**105**(12):2193-2196

[52] Zadok D, Tran DB, Twa M, Carpenter M, Schanzlin DJ. Pneumotonometry versus Goldmann tonometry after laser in situ keratomileusis for myopia. Journal of Cataract and Refractive Surgery. 1999;**25**(10):1344-1348

[53] Tonnu PA, Ho T, Sharma K, White E, Bunce C, Garway-Heath D. A comparison of four methods of tonometry: Method agreement and interobserver variability. The British Journal of Ophthalmology. 2005;**89**(7):847-850

[54] Tonnu PA, Ho T, Newson T, El Sheikh A, Sharma K, White E, et al. The influence of central corneal thickness and age on intraocular pressure measured by pneumotonometry, non-contact tonometry, the Tono-pen XL, and Goldmann applanation tonometry. The British Journal of Ophthalmology. 2005;**89**(7):851-854

[55] De Moraes CGV, Prata TS, Liebmann J, Ritch R. Modalities of tonometry and their accuracy with respect to corneal thickness and irregularities. Journal of Optometry. 2008;**1**(2):43-49

[56] Scibilia GD, Ehlers WH, Donshik PC. The effects of therapeutic contact lenses on intraocular pressure measurement. The CLAO Journal. 1996;**22**(4):262-265

[57] Frenkel RE, Hong YJ, Shin DH. Comparison of the Tono-Pen to the Goldmann applanation tonometer. Archives of Ophthalmology. 1988;**106**(6):750-753

[58] Badakere SV, Chary R, Choudhari NS, Rao HL, Garudadri C, Senthil S. Agreement of intraocular pressure measurement of Icare ic200 with Goldmann Applanation tonometer in adult eyes with normal cornea. Ophthalmology Glaucoma. 2021;**4**(1):89-94

[59] Brusini P, Salvetat ML, Zeppieri M, Tosoni C, Parisi L. Comparison of ICare tonometer with Goldmann applanation

tonometer in glaucoma patients. Journal of Glaucoma. 2006;**15**(3):213-217

[60] Gao F, Liu X, Zhao Q, Pan Y. Comparison of the iCare rebound tonometer and the Goldmann applanation tonometer. Experimental and Therapeutic Medicine. 2017;**13**(5):1912-1916

[61] Nakakura S, Mori E, Fujio Y, Fujisawa Y, Matsuya K, Kobayashi Y, et al. Comparison of the intraocular pressure measured using the new rebound tonometer Icare ic100 and Icare TA01i or Goldmann Applanation tonometer. Journal of Glaucoma. 2019;**28**(2):172-177

[62] Liu J, De Francesco T, Schlenker M, Ahmed II. Icare home tonometer: A review of characteristics and clinical utility. Clinical Ophthalmology. 2020;**14**:4031-4045

[63] Chen M, Zhang L, Xu J, Chen X, Gu Y, Ren Y, et al. Comparability of three intraocular pressure measurement: iCare pro rebound, non-contact and Goldmann applanation tonometry in different IOP group. BMC Ophthalmology. 2019;**19**(1):225

[64] Nakamura M, Darhad U, Tatsumi Y, Fujioka M, Kusuhara A, Maeda H, et al. Agreement of rebound tonometer in measuring intraocular pressure with three types of applanation tonometers. American Journal of Ophthalmology. 2006;**142**(2):332-334

[65] Jose J, Ve RS, Pai HV, Biswas S, Parimi V, Poojary P, et al. Agreement and repeatability of Icare ic100 tonometer. Indian Journal of Ophthalmology. 2020;**68**(10):2122-2125

[66] Nakakura S, Asaoka R, Terao E, Nagata Y, Fukuma Y, Oogi S, et al. Evaluation of rebound tonometer iCare IC200 as compared with IcarePRO and Goldmann applanation tonometer in patients with glaucoma. Eye Vision (Lond). 2021;**8**(1):25

[67] Berg C, Doniger SJ, Zaia B, Williams SR. Change in intraocular pressure during point-of-care ultrasound. The Western Journal of Emergency Medicine. 2015;**16**(2):263-268

[68] Chakraborty AK, Majumder M, Sen S. Comparison of transpalpebral tonometer with Goldmann applanation tonometer. Taiwan Journal of Ophthalmology. 2014;**4**(3):110-115

[69] Shemesh G, Man O, Michaeli A, Varssano D, Lazar M. Pressure phosphene tonometry versus Goldmann Applanation tonometry for measuring intraocular pressure before and after LASIK. Journal of Refractive Surgery. 2007;**23**(4):405-409

[70] Alvarez T. The proview phosphene tonometer fails to measure ocular pressure accurately in clinical practice*1. Ophthalmology. 2004;**111**(6):1077-1085

[71] Danesh-Meyer HV, Niederer R, Gaskin BJ, Gamble G. Comparison of the Proview pressure phosphene tonometer performed by the patient and examiner with the Goldmann applanation tonometer. Clinical & Experimental Ophthalmology. 2004;**32**(1):29-32

[72] Kaufmann C, Bachmann LM, Thiel MA. Intraocular pressure measurements using dynamic contour tonometry after laser in situ keratomileusis. Investigative Ophthalmology & Visual Science. 2003;**44**(9):3790-3794

[73] Punjabi OS, Kniestedt C, Stamper RL, Lin SC. Dynamic contour tonometry: Principle and use. Clinical

& Experimental Ophthalmology. 2006;**34**(9):837-840

[74] Katsimpris JM, Theoulakis PE, Vasilopoulos K, Skourtis G, Papadopoulos GE, Petropoulos IK. Correlation between central corneal thickness and intraocular pressure measured by Goldmann Applanation tonometry or Pascal dynamic contour tonometry. Klinische Monatsblätter für Augenheilkunde. 2015;**232**(4):414-418

[75] Siganos DS, Papastergiou GI, Moedas C. Assessment of the Pascal dynamic contour tonometer in monitoring intraocular pressure in unoperated eyes and eyes after LASIK. Journal of Cataract and Refractive Surgery. 2004;**30**(4):746-751

[76] Lee SY, Kim EW, Choi W, Park CK, Kim S, Bae HW, et al. Significance of dynamic contour tonometry in evaluation of progression of glaucoma in patients with a history of laser refractive surgery. The British Journal of Ophthalmology. 2020;**104**(2):276-281

[77] Kandarakis A, Soumplis V, Pitsas C, Kandarakis S, Halikias J, Karagiannis D. Comparison of dynamic contour tonometry and Goldmann Applanation tonometry following penetrating keratoplasty. Canadian Journal of Ophthalmology. 2010;**45**(5):489-493

[78] Schwenn O, Troost R, Vogel A, Grus F, Beck S, Pfeiffer N. Ocular pulse amplitude in patients with open angle glaucoma, normal tension glaucoma, and ocular hypertension. The British Journal of Ophthalmology. 2002;**86**(9):981-984

[79] Fogagnolo P, Figus M, Frezzotti P, Iester M, Oddone F, Zeppieri M, et al. Test-retest variability of intraocular pressure and ocular pulse amplitude for dynamic contour tonometry: A multicentre study. The British Journal of Ophthalmology. 2010;**94**(4):419-423

[80] Wang AS, Alencar LM, Weinreb RN, Tafreshi A, Deokule S, Vizzeri G, et al. Repeatability and reproducibility of Goldmann applanation, dynamic contour, and ocular response analyzer tonometry. Journal of Glaucoma. 2013;**22**(2):127-132

[81] Francis BA, Hsieh A, Lai MY, Chopra V, Pena F, Azen S, et al. Effects of corneal thickness, corneal curvature, and intraocular pressure level on Goldmann applanation tonometry and dynamic contour tonometry. Ophthalmology. 2007;**114**(1):20-26

[82] Kim SH, Lee EJ, Han JC, Sohn SW, Rhee T, Kee C. The effect of diurnal fluctuation in intraocular pressure on the evaluation of risk factors of progression in Normal tension Glaucoma. PLoS One. 2016;**11**(10):e0164876

[83] Hasegawa K, Ishida K, Sawada A, Kawase K, Yamamoto T. Diurnal variation of intraocular pressure in suspected normal-tension glaucoma. Japanese Journal of Ophthalmology. 2006;**50**(5):449-454

[84] Renard E, Palombi K, Gronfier C, Pepin JL, Noel C, Chiquet C, et al. Twenty-four hour (Nyctohemeral) rhythm of intraocular pressure and ocular perfusion pressure in normal-tension glaucoma. Investigative Ophthalmology & Visual Science. 2010;**51**(2):882-889

[85] De Moraes CG, Jasien JV, Simon-Zoula S, Liebmann JM, Ritch R. Visual field change and 24-hour IOP-related profile with a contact Lens sensor in treated Glaucoma patients. Ophthalmology. 2016;**123**(4):744-753

[86] Leonardi M, Pitchon EM, Bertsch A, Renaud P, Mermoud A. Wireless contact

lens sensor for intraocular pressure monitoring: Assessment on enucleated pig eyes. Acta Ophthalmologica. 2009;**87**(4):433-437

[87] Mansouri K, Weinreb RN, Liu JHK. Efficacy of a contact lens sensor for monitoring 24-H intraocular pressure related patterns. Lin H, editor. PLoS One. 2015;**10**(5):e0125530

[88] Mansouri K, Medeiros FA, Tafreshi A, Weinreb RN. Continuous 24-hour monitoring of intraocular pressure patterns with a contact lens sensor: Safety, tolerability, and reproducibility in patients with glaucoma. Archives of Ophthalmology. 2012;**130**(12):1534-1539

[89] Holló G, Kóthy P, Vargha P. Evaluation of continuous 24-hour intraocular pressure monitoring for assessment of prostaglandin-induced pressure reduction in glaucoma. Journal of Glaucoma. 2014;**23**(1):e6-e12

[90] Lorenz K, Korb C, Herzog N, Vetter JM, Elflein H, Keilani MM, et al. Tolerability of 24-hour intraocular pressure monitoring of a pressure-sensitive contact lens. Journal of Glaucoma. 2013;**22**(4):311-316

[91] Tojo N, Abe S, Ishida M, Yagou T, Hayashi A. The fluctuation of intraocular pressure measured by a contact Lens sensor in Normal-tension Glaucoma patients and nonglaucoma subjects. Journal of Glaucoma. 2017;**26**(3):195-200

[92] Choritz L, Mansouri K, Van Den Bosch J, Weigel M, Dick HB, Wagner M, et al. Telemetric measurement of intraocular pressure via an implantable pressure sensor—12-month results from the ARGOS-02 trial. American Journal of Ophthalmology. 2020;**209**:187-196

[93] Szurman P, Gillmann K, Seuthe AM, Dick HB, Hoffmann EM, Mermoud A, et al. EYEMATE-SC Trial: Twelve-month safety, performance, and accuracy of a Suprachoroidal sensor for telemetric measurement of intraocular pressure. Ophthalmology. 2023;**130**(3):304-312

[94] Melki S, Todani A, Cherfan G. An implantable intraocular pressure transducer: Initial safety outcomes. JAMA Ophthalmology. 2014;**132**(10):1221-1225

[95] Koutsonas A, Walter P, Roessler G, Plange N. Implantation of a novel telemetric intraocular pressure sensor in patients with glaucoma (ARGOS study): 1-year results. Investigative Ophthalmology & Visual Science. 2015;**56**(2):1063-1069

[96] Mariacher S, Ebner M, Januschowski K, Hurst J, Schnichels S, Szurman P. Investigation of a novel implantable suprachoroidal pressure transducer for telemetric intraocular pressure monitoring. Experimental Eye Research. 2016;**151**:54-60

[97] Szurman P, Mansouri K, Dick HB, Mermoud A, Hoffmann EM, Mackert M, et al. Safety and performance of a suprachoroidal sensor for telemetric measurement of intraocular pressure in the EYEMATE-SC trial. The British Journal of Ophthalmology. 2023;**107**(4):518-524

[98] Meyenberg A, Iliev ME, Eschmann R, Frueh BE. Dynamic contour tonometry in keratoconus and postkeratoplasty eyes. Cornea. 2008;**27**(3):305-310

[99] Injectsense. [Internet] 2023. Available from https://www.injectsense.com/technology [Accessed August 15, 2023]

Chapter 7

When to Treat Ocular Hypertension?

Gian Franco Díez Cattini

Abstract

How to decide which patient should receive treatment to prevent conversion to glaucoma? Ocular hypertension is the only treatable risk factor for glaucoma, yet not all patients with hypertension develop glaucoma, and not all glaucoma patients have ocular hypertension. Deciding when and how to start treatment can be challenging, particularly in patients without other risk factors. When facing this dilemma one should ponder: the patient's risk tolerance/avoidance, the prospect of long-term topical treatment including adverse effects, the psychological and economical burden of a pre-disease state and its treatment, treatment compliance, and the possibility of over-treating. New tools such as AI-guided image analysis, improved testing algorithms, and novel minimally invasive treatments may help control and weigh this risk factor more conveniently, avoiding over-treatment but preventing glaucoma-related blindness.

Keywords: ocular hypertension, treatment, topical medication, risk factors, laser in glaucoma

1. Introduction

Ocular hypertension is the most recognizable risk factor for glaucoma-related vision loss and has long been associated with an increased risk of blindness. Our understanding of the pathophysiology of ocular hypertension has significantly changed in recent decades, and new clinical evidence has allowed us to fine-tune guidelines regarding when and how to treat ocular hypertension in order to avoid vision loss.

Intraocular pressure (IOP) is the result of the dynamic process of aqueous inflow and outflow from the anterior segment of the eye [1]. In the healthy eye, the aqueous balance is fairly stable. Inflow involves a highly specialized active secretion process carried out in the ciliary body epithelium, while outflow is influenced by biochemical and mechanical factors (pressure-dependent and pressure-independent mechanisms) that facilitate drainage. Glaucoma and ocular hypertension are typically caused by a restriction in the outflow of fluid from the eye, either in the trabecular outflow apparatus (Schlemm's canal, trabecular meshwork), the uveoscleral pathway, or the post-trabecular structures [2, 3].

Several high-quality randomized clinical trials have demonstrated that lowering IOP slows glaucoma conversion, glaucoma progression, and outcomes related to visual loss [4–9]. However, not all eyes respond equally to the same amount of IOP elevation: most hypertensive eyes do not develop glaucoma even without treatment, and a sizeable proportion of glaucomatous eyes never register high IOPs (normal tension glaucoma) but respond well to IOP-lowering therapies [4]. These differences can be attributed to varying IOP threshold sensitivities among patients or to different mechanisms of retinal nerve fiber layer damage (vascular dysregulation, trans-lamina cribrosa pressure gradient).

Treating ocular hypertension is therefore not straightforward, and an individualized approach considering the magnitude and mechanism of intraocular pressure elevation, the daily pressure fluctuations, the presence of other risk factors (positive family history, age, low-corneal hysteresis), the patient's risk tolerance, and the safety of available treatment options must be adopted.

2. Risk factors for ocular hypertension

Ocular hypertension (OHT) is defined when finding high-intraocular pressure readings (above 21 mmHg) in a patient without evidence of glaucomatous neuropathy. The cut-off value of "normal" IOP is based on epidemiological studies in a healthy population and is not representative of all ethnicities or age groups but serves as an accurate guideline to which all patients are compared.

The most relevant risk factors for the development of ocular hypertension are:

1. Ethnicity and age: Afro-descendants and Latinos have higher average IOPs than Caucasians. These populations also have higher prevalence of glaucoma and are considered at higher risk of glaucoma-related blindness [9]. Intraocular pressure is higher in older populations, and glaucoma prevalence rises through life in a linear fashion.

2. Family history: The presence of a first-degree relative with ocular hypertension or glaucoma increases the risk of developing ocular hypertension three to four times greater [10]. Multiple genetic loci have been identified, but inheritance is considered multifactorial with variable expressivity [11]. When there is a clear family history, hypertension tends to present itself sooner and in higher magnitude in subsequent generations.

3. Ocular pathology: Some ocular abnormalities increase the odds of developing high IOP; the most common are: pseudoexfoliation syndrome, pigment dispersion, history of active inflammation, trauma, or previous surgery [12]. Secondary OHT and glaucoma are usually approached differently from cases with no identifiable cause and management can differ. In this chapter, we will be referring almost exclusively to primary OHT and primary open-angle glaucoma (POAG).

4. Central corneal thickness and corneal hysteresis: Central corneal thickness has been found to be an independent risk factor for the development of glaucoma [7]. The theory of this association has not yet been proven but involves the relationship of corneal thickness and corneal force dampening with ocular rigidity and lamina cribrosa deformity.

Some risk factors that are frequently mentioned in literature on glaucoma and OHT can be also considered early manifestations of the disease and even be the only finding in such cases and should therefore be considered when calculating the overall risk of progression. These are:

1. Cup-to-disk ratio (CDR): A larger cup can be found in normal individuals, particularly in large discs. The cup-disk ratio is usually the first finding that alerts the clinician to look for possible glaucomatous retinal nerve fiber layer loss. In at-risk individuals, a larger CDR increases the risk of lifetime conversion to glaucoma.

2. Disc hemorrhages: also called splinter hemorrhages, are commonly found in glaucoma patients and are considered hallmark of glaucomatous progression. The finding of disc hemorrhages can weigh considerably when assessing overall risk.

3. Peripapillary atrophy: Atrophy around the optic disc is common in glaucoma and may sometimes be a predictor of RFNL thinning. Beta peripapillary atrophy is associated with higher probability of glaucoma.

3. What is the evidence on treating ocular hypertension?

3.1 Early treatment lowers the risk of glaucoma conversion in OHT

The Ocular Hypertension Treatment Study (OHTS) conducted a randomized trial with 1408 patients having ocular hypertension (defined as IOP above 24 mmHg in at least one eye) assigned to either treatment or observation [6, 7]. After a 60-month follow-up, only 4.4% of patients in the medication group and 9.5% in the observation group developed Primary Open-Angle Glaucoma (POAG), illustrating a remarkable 60% risk reduction. Over 13 years, the cumulative proportion of patients developing POAG in the treatment group was 0.16, compared to 0.22 in the observation group [6].

It is important to notice that most of the observation cohort did not develop POAG. Individual risk factors should be taken into consideration when starting medical therapy to avoid over-treatment.

3.2 Higher risk patients have greater odds for glaucoma conversion in OHT

The OHTS study also revealed that African American participants showed a higher proportion of POAG development in both groups, with 6.9% in the medication group and 12.7% in the observation group [7]. A risk factor analysis was carried out, finding that those associated with a higher risk of POAG were family history of glaucoma, corneal thickness of 555 μm or less, older age, higher IOP, and larger cup-to-disk ratio.

In the 12 year follow-up phase, early treatment was found to be more protective than delayed treatment (started after the first 7.5 years of the first stage study), particularly in the higher risk group of patients. Higher risk individuals should be approached more closely, and the decision of early treatment be considered sooner, deferring treatment in these patients may carry a penalty that may not be caught up if treatment is started only after finding glaucomatous damage.

3.3 IOP-lowering treatment slows glaucomatous progression

The Early Manifest Glaucoma Trial (EMGT) examined the efficacy of reducing IOP to prevent the progression of POAG. In this study, 255 patients with newly diagnosed POAG were randomized into two groups: a treatment group receiving laser trabeculoplasty and hypotensive drops, and an observation group. Patients in the observation group exhibited greater progression in visual fields and optic disc findings, accounting for 62% compared to 45% in the treatment group. Although both groups showed progression over time, the control group experienced more significant deterioration, suggesting the beneficial effect of treatment [8].

In established advanced glaucoma, the Advanced Glaucoma Intervention Study (AGIS) found that eyes with average IOP greater than 17.5 mmHg progressed more than those under 14 mmHg, and patients that had IOP under 18 mmHg on all visits in the 6-year follow-up had close to no change in the basal visual field defects [9].

3.4 IOP-lowering therapy slows normal tension glaucoma

IOP-lowering therapy also slows down the progression of Normal Tension Glaucoma. The Collaborative Normal Tension Glaucoma Study enrolled 140 patients and randomly assigned them to two groups [13]. One group received IOP reduction of 30% from baseline, while the other remained untreated (control group). The study found that reducing IOP by 30% slowed the rate of progression in visual fields compared to the control group. Notably, up to 65% of patients in the control group showed no progression during the 5-year follow-up.

4. Treatment strategies

Traditionally, IOP treatment starts with topical hypotensive medications, either in ocular hypertension or glaucomatous patients. These topical medications are typically the first-line approach due to their ease of use, effectiveness in lowering IOP, and relatively low risk of systemic side effects. Most glaucoma patients, and many OHT patients will use one or more topical hypotensive medication in their lifetime. Careful selection of an appropriate medication can help control the disease with great integration to the patient's daily routines. This integration facilitates a consistent management plan, increasing the likelihood of treatment adherence and enhancing the overall efficacy of the therapeutic regimen.

Commonly prescribed topical medications include prostaglandin analogs which act enhancing uveoscleral outflow; beta-blockers, alpha-agonists, and carbonic anhydrase inhibitors all act as aqueous production inhibitors, and rho kinase inhibitors, a newer class of hypotensive medications which act on the trabecular meshwork. Through these diverse mechanisms, they lower the IOP and halt the degeneration of retinal ganglion cells.

Recent efforts have been made to decrease medication burden and adverse effects, increase adherence, persistence, and overall patient quality of life and other patient-reported outcomes. Out of these efforts, preservative-free and fixed-combination medications have entered the market and the use of other therapies such as selective laser trabeculoplasty (SLT) and minimally invasive glaucoma surgery (MIGS) has claimed a place in the treatment of early stages of glaucoma and ocular hypertension.

This novel point of view is encapsulated in the term "interventional glaucoma," signifying a shift in the approach toward pre-emptive tactics aimed at stalling glaucomatous progression and safeguarding vision. Central to this approach are early and safer interventions designed to strengthen adherence, amplify glaucoma control, and gauge the enhancements in quality of life. By embracing this perspective, the need for medication escalation and medication-linked adversities is decreased, while also reducing the fluctuations of intraocular pressure and the overarching trajectory of disease progression.

4.1 Medical therapy

Whenever possible, medical therapy should start with a single agent; drug combinations can be considered when a greater amount of IOP decrease is desired, and it is unlikely to be achieved with a single medication.

Prostaglandin analogues (PGAs) are usually the first choice as they have the highest potency, once-daily administration and a good safety profile [14]. There is increasing availability of preservative-free formulas worldwide. Special care should be taken in patients with history of ocular inflammation or complicated cataract surgery; otherwise, adverse effects are usually limited to mild ocular redness and discomfort. Prostaglandin orbitopathy (periorbital fat atrophy) and other cosmetic changes (hyperpigmentation) can become significant and should be assessed at every visit.

Beta-blockers are also considered a good first-line choice when PGAs are not ideal or as adjunctive therapy when treatment escalation is needed. Beta-blockers should be used with care or avoided in patients with pulmonary of cardiac problems. Other medications are usually reserved as third-line agents either alone or in fixed combinations when IOP is not at target or progression is documented. Newer drug classes such as rho-kinase inhibitors (netarsudil) and nitrous oxide donors (latanoprostene bunod) are usually used in combination with PGAs.

4.2 Selective laser trabeculoplasty (SLT)

Laser Trabeculoplasty has been utilized for glaucoma treatment for over two decades [15]. However, recent evidence from randomized controlled studies has proven its efficacy as a first-line therapy in patients with ocular hypertension (OHT) and varying degrees of glaucoma.

The LIGHT study (Laser in Glaucoma and Ocular Hypertension Treatment study) compared selective laser trabeculoplasty (SLT) to eye drops as the primary treatment. It assessed various outcomes, including quality of life, cost-effectiveness, intraocular pressure (IOP) reduction efficacy, and adverse effects. At the 36-month mark, no difference in quality-of-life scores was observed. However, there was a slight advantage in terms of the percentage of eyes reaching the target IOP, fewer treatment escalations, less progression in visual fields, and a reduced need for surgery. During the 6-year extension, 69.8% of patients in the SLT group maintained their target IOP without requiring medical treatment [16, 17].

SLT can be employed at different stages of the disease, but it exhibits a more favorable outcome profile in cases of ocular hypertension and mild primary open-angle glaucoma [16]. This approach should be considered for all patients with open-angle glaucoma or OHT in whom the decision to initiate treatment has been made. SLT can potentially defer the need for medication and mitigate associated adverse effects for a substantial portion of patients.

4.3 Glaucoma surgery

Many new surgical procedures for glaucoma management have appeared in recent years. Novel devices and techniques approaching the trabecular meshwork, the suprachoroidal space or subconjunctival filtration with improved safety profiles, quick recovery and good efficacy are now available for the treatment of early to severe glaucoma [18, 19]. These procedures are encompassed in the ever-expanding umbrella of MIGS (Minimally Invasive Glaucoma Surgery).

MIGS was initially designed for mild to moderate glaucoma, to be used in combination with cataract extraction. With widespread use and more information on results, they have been found to be effective also in more severe cases or as standalone therapy.

Trabeculectomy and glaucoma drainage devices are traditional incisional surgeries used for uncontrolled ocular hypertension and glaucoma. They are left as a secondary choice when other therapies (medication, laser, or MIGS) are insufficient or there is need for a single-digit target IOP [19]. Although they provide the lowest target IOP, postoperative complications and reinterventions are frequent, and recovery is slow. They continue to be very valuable tools for the management of refractory glaucoma and many cases of secondary glaucoma.

5. What to consider when starting treatment?

5.1 Age

The age of patients is important not only as a risk factor but also in terms of treatment considerations. When deciding on the timing and method of treatment, the patient's life expectancy at the time of diagnosis should be carefully taken into account. While most topical medications are safe and well-tolerated and are still the first line of treatment, the prospect of undergoing treatment for decades can be discouraging.

Studies have shown that long-term use of topical treatments can lead to a decrease in conjunctival goblet cell density, reduced tear stability, and the development of ocular surface disease. Additionally, the prolonged use of prostaglandin analogs can result in cosmetic changes such as periorbital fat atrophy, iris color alteration, and eyelid pigmentation. Alpha agonists may lose some of their effectiveness due to tachyphylaxis and can even cause delayed-onset allergic conjunctivitis. It is worth noting that all hypotensive medications are associated with higher rates of cataract surgery.

When feasible, delaying the initiation of ocular hypotensive medications in younger individuals is a reasonable approach, especially in cases of low-risk ocular hypertension patients. For some patients with evident glaucomatous findings, alternative treatments like selective laser trabeculoplasty (SLT) can be used as they have demonstrated comparable efficacy without the need for a drop regimen. Trabecular minimally invasive glaucoma surgeries (MIGS) or even trabeculectomy might yield excellent long-term outcomes in cases of juvenile primary open-angle glaucoma (POAG) and considerably reduce the overall medication burden over time.

Conversely, in the case of older patients, a less aggressive approach might be considered if the glaucoma is mild and stable, if their life expectancy is shorter, or if they have significant and life-threatening systemic comorbidities. In such instances, it becomes crucial to lower intraocular pressure (IOP) sufficiently to prevent substantial vision loss within the expected lifespan. The chosen treatment should possess the best available safety, efficacy, and tolerability profile. Surgery may be preferred in cases where compliance with a drop-based therapy is difficult or unreliable.

5.2 Ocular surface disease

Ocular surface disease (OSD) is a highly prevalent ophthalmic condition. Dry eye symptoms are experienced by 8 to 30% of the general population, with an increase to 59% among glaucoma patients [20–22]. This notable increase can be attributed to the prolonged exposure to topical treatments, many of which contain benzalkonium chloride (BAK), a common preservative known to disrupt tear film balance and trigger inflammation within various ocular structures, especially the cornea [23–26].

In the initiation of ocular hypotensive medications, a meticulous assessment of the ocular surface is important. This evaluation should encompass pre-existing conditions such as blepharitis, reduced tear break-up time, conjunctival and corneal epithelial defects. Additionally, it is helpful to objectively measure patient symptoms using standardized questionnaires like the Ocular Surface Disease Index (OSDI) [27]. This information helps select the best suited treatment: drop-free therapies such as SLT in patients with existent OSD or preservative-free medications and simpler, once-daily drop regimens for patients at risk.

OSD should be treated as soon as diagnosed. Eyelid margin hygiene and preservative-free lubricant drops should be prescribed to all patients with OSD and glaucoma. Ocular surface immune-modulating agents such as cyclosporine can help minimize chronic inflammatory changes to the ocular surface. Other more intensive therapies like punctual plugs, 20% autologous serum drops, or intense pulsed light should be considered if necessary.

5.3 Glaucoma severity

The amount of visual field remaining and at risk is a good surrogate measure of how intense treatment should be. Visual field loss is dependent on many mechanisms but is directly proportional to the time and magnitude of ocular hypertension. As a general rule, severe visual loss should be treated aggressively and monitored closely to avoid visual disability.

It has also been demonstrated that a target IOP in the lower teens, or even in single digits, stops progression in advanced glaucoma [9]. In these cases, it is safest to avoid IOP spikes during the day. Treatment choice should take daily fluctuations into account, considering some medications have less efficacy during the night [28].

If visual disability has occurred as a result of other ocular diseases, such as macular degeneration or diabetic retinopathy, it is also wise to have a low threshold for starting treatment when ocular hypertension is diagnosed.

5.4 Patient preference

Clear and sufficient communication is pivotal in establishing a trusting relationship between patients and their care providers. Effective communication yields positive impacts on both clinical outcomes and patient-related results. It promotes adherence to treatment regimens and cultivates realistic expectations regarding the disease. A thorough understanding of potential risks and diverse treatment options empowers patients to make informed decisions, even when those decisions involve taking calculated risks.

Risk-averse patients might opt to start medication even if their glaucoma risk is considered low. In contrast, others might feel at ease with an observation-based approach, viewing it as a means to postpone treatment and the potential complications that come with it. Equipping patients with the knowledge of what to anticipate in each scenario facilitates clinical decision-making, even when adverse outcomes do arise.

As ophthalmologists, it is crucial to provide guidance grounded in up-to-date evidence toward the most suitable choice for the patient's specific issue. By offering an informed insight and assisting patients in navigating their options, clinicians ensure that decisions are well-informed and aligned with the individual's needs.

5.5 Overall risk of progression

The significance of intraocular pressure (IOP) treatment lies within its integration into a comprehensive therapeutic strategy aimed at preserving visual function. Visual function encompasses vital aspects like visual acuity, depth perception, and the extent of the visual field. It holds immense importance in the overall quality of life for each individual patient. Consequently, the fixation of a target IOP only constitutes one facet of treatment. This fixation becomes adjunctive to the observation of functional or structural damage and its rate of change over time [29].

The rate of progression serves as a crucial indicator of disease dynamics. It offers insights into the speed of deterioration, the effectiveness of therapeutic interventions, and empowers clinicians to gauge the likelihood of visual impairment within specific timeframes. To compute this rate, patients need to undergo frequent visual field and structural optical coherence tomography (OCT) tests. While these tests can be cumbersome, they are necessary to establish baseline measurements, evaluate inter-test variability, and ascertain a trend of worsening.

In cases of ocular hypertension, notable thinning of the retinal nerve fiber layer (RFNL) over time might signal the earliest indication of glaucomatous neuropathy and the necessity for intervention. For patients already diagnosed with glaucoma, a faster than expected progression rate should prompt a reevaluation of the ongoing treatment plan. This could involve surgical interventions, medication escalation, and a thorough assessment of associated systemic factors or overlooked instances of intraocular pressure spikes, as well as adherence issues.

When visual function and its quantifiable parameters remain stable over time, the chosen approach, be it observation or an active treatment regimen, is likely well suited for the patient's current condition. In such cases, ongoing monitoring can continue without needing further alterations.

6. General guidelines

6.1 Is there evidence of glaucomatous neuropathy or high risk for glaucoma development?

If the answer is yes, IOP-lowering treatment is warranted. Consider different options and comment with the patient the benefits and caveats of every approach. Evaluate the ocular surface and presence of ocular and systemic comorbidities (cataract, retinal disease, asthma, cardiac disease).

When there are no signs of glaucomatous neuropathy or high-risk factors and the patient is comfortable with an observation-only approach, educate on the importance of follow-up and continue frequent visual field and OCT testing in order to detect a worsening trend or the appearance of signs of high risk (disc hemorrhages, RFNL defects, increasing IOP).

6.2 Is there ocular surface disease?

Pre-existing OSD should be managed and added hypotensive treatments should be chosen considering the best option for both IOP management and the potential for OSD worsening. SLT should be considered the first-line treatment if possible, if additional IOP lowering is needed or SLT not possible, preservative-free and fixed combinations are the better choice.

Limiting the amount of daily drops and avoiding BAK-related toxicity can go a long way to ensure compliance to treatment and happy patients. Motivated patients are more adherent to accorded treatments. When using topical medications, it is better to start with monotherapy and adding/changing agents if target IOP is not met or adverse reactions appear.

6.3 Is there concomitant cataract?

Consider cataract surgery combined with an IOP-lowering procedure. Cataract surgery on itself has IOP-lowering effect on most patients, adding MIGS procedures/devices can lower IOP even further. Glaucoma incisional surgery can be a good choice if other options are unavailable or a larger IOP drop is needed [29].

7. Conclusions

IOP-lowering therapy should be tailored to each patient considering important risk factors (magnitude of OHT, family history, corneal thickness and hysteresis, viable visual function) and comorbidities (ocular surface disease, cataract, retinopathy), and potential problems with adherence.

Acknowledgements

I would like to thank Lucía Ximena Sadurní González MD.

Conflict of interest

The authors declare no conflict of interest.

Author details

Gian Franco Díez Cattini
Hospital de la Luz, Mexico City, Mexico

*Address all correspondence to: gdiezcattini@gmail.com

References

[1] Toris CB, Koepsell SA, Yablonski ME, Camras CB. Aqueous humor dynamics in ocular hypertensive patients. Journal of Glaucoma. 2002;**11**:253-258

[2] Bill A, Phillips CI. Uveoscleral drainage of aqueous humour in human eyes. Experimental Eye Research. 1971;**12**:275-281

[3] Bill A. Uveoscleral drainage of aqueous humor: Physiology and pharmacology. Progress in Clinical and Biological Research. 1989;**312**:417-427

[4] Anderson DR, Drance SM, Schulzer M. Factors that predict the benefit of lowering intraocular pressure in normal tension glaucoma. American Journal of Ophthalmology. 2003;**136**:820-829

[5] Gordon MO, Kass MA. The ocular hypertension treatment study: Design and baseline description of the participants. Archives of Ophthalmology. 1999;**117**:573-583

[6] Kass MA, Heuer DK, Higginbotham EJ, et al. The ocular hypertension treatment study: A randomized trial determines that topical ocular hypotensive medication delays or prevents the onset of primary open-angle glaucoma. Archives of Ophthalmology. 2002;**120**:701-713. discussion 829-730

[7] Gordon MO, Beiser JA, Brandt JD, et al. The ocular hypertension treatment study: Baseline factors that predict the onset of primary open-angle glaucoma. Archives of Ophthalmology. 2002;**120**:714-720. discussion 829-730

[8] Leske MC, Heijl A, Hyman L, Bengtsson B. Early manifest glaucoma trial: Design and baseline data. Ophthalmology. 1999;**106**:2144-2153

[9] The advanced glaucoma intervention study (AGIS): 7. The relationship between control of intraocular pressure and visual field deterioration. The AGIS investigators. American Journal of Ophthalmology. 2000;**130**:429-440

[10] Tielsch JM, Katz J, Sommer A, Quigley HA, Javitt JC. Family history and risk of primary open angle glaucoma. The Baltimore eye survey. Archives of Ophthalmology. 1994;**112**(1):69-73

[11] Khan AO. Genetics of primary glaucoma. Current Opinion in Ophthalmology. 2011;**22**(5):347-355

[12] Krishnadas R, Ramakrishnan R. Secondary glaucomas: The tasks ahead. Community Eye Health. 2001;**14**(39):40-42

[13] Anderson DR, Normal Tension Glaucoma Study. Collaborative normal tension glaucoma study. Current Opinion in Ophthalmology. 2003;**14**(2):86-90

[14] Wu X, Yang X, Liang Q, Xue X, Huang J, Wang J, et al. Drugs for the treatment of glaucoma: Targets, structure-activity relationships and clinical research. European Journal of Medicinal Chemistry. 2021;**226**:113842

[15] Katz LJ, Steinmann WC, Kabir A, Molineaux J, Wizov SS, Marcellino G. Selective laser trabeculoplasty versus medical therapy as initial treatment of glaucoma: A prospective, randomized trial. Journal of Glaucoma. 2012;**21**(7):460-468

[16] Gazzard G, Konstantakopoulou E, Garway-Heath D, et al. Laser in glaucoma and ocular hypertension (LiGHT) trial. A multicentre, randomised controlled

trial: Design and methodology. The British Journal of Ophthalmology. 2018;**102**(5):593-598

[17] Gazzard G, Konstantakopoulou E, Garway-Heath D, Adeleke M, Vickerstaff V, Ambler G, et al. LiGHT trial: 6-year results of primary selective laser trabeculoplasty versus eye drops for the treatment of glaucoma and ocular hypertension. Ophthalmology. 2022;**130**:139-151

[18] Pereira ICF, van de Wijdeven R, Wyss HM, Beckers HJM, den Toonder JMJ. Conventional glaucoma implants and the new MIGS devices: A comprehensive review of current options and future directions. Eye (London, England). 2021;**35**(12):3202-3221

[19] Lim R. The surgical management of glaucoma: A review. Clinical & Experimental Ophthalmology. 2022;**50**(2):213-231

[20] Chan CC, Crowston JG, Tan R, et al. Ocular surface disease in patients with glaucoma from Australia. Asia Pacific Journal of Ophthalmology. 2013;**2**:79-87

[21] Fechtner RD, Godfrey DG, Budenz D, et al. Prevalence of ocular surface complaints in patients with glaucoma using topical intraocular pressure-lowering medications. Cornea. 2010;**29**:618-621

[22] Ruangvaravate N, Prabhasawat P, Vachirasakchai V, Tantimala R. High prevalence of ocular surface disease among glaucoma patients in Thailand. Journal of Ocular Pharmacology and Therapeutics. 2018;**34**:387-394

[23] Baudouin C. Detrimental effect of preservatives in eye drops: Implications for the treatment of glaucoma. Acta Ophthalmologica. 2008;**86**:716-726

[24] Vitoux MA, Kessal K, Melik Parsadaniantz S, et al. Benzalkonium chloride-induced direct and indirect toxicity on corneal epithelial and trigeminal neuronal cells: Proinflammatory and apoptotic responses *in vitro*. Toxicology Letters. 2020;**319**:74-84

[25] Zhu W, Kong X, Xu J, Sun X. Effects of long-term antiglaucoma eye drops on conjunctival structures: An in vivo confocal microscopy study. Journal of Ophthalmology. 2015;**2015**:165475

[26] Pisella PJ, Pouliquen P, Baudouin C. Prevalence of ocular symptoms and signs with preserved and preservative free glaucoma medication. The British Journal of Ophthalmology. 2002;**86**:418-423

[27] Portela RC, Fares NT, Machado LF, et al. Evaluation of ocular surface disease in patients with glaucoma: Clinical parameters, self-report assessment, and keratograph analysis. Journal of Glaucoma. 2018;**27**:794-801

[28] Stewart WC, Konstas AGP, Nelson LA, Kruft B. Meta-analysis of 24-hour intraocular pressure studies evaluating the efficacy of glaucoma medicines. Ophthalmology. 2008;**115**:1117-1122.e1

[29] Saunders LJ, Medeiros FA, Weinreb RN, Zangwill LM. What rates of glaucoma progression are clinically significant? Expert Reviews in Ophthalmology. 2016;**11**(3):227-234

Chapter 8

Pseudo-Exfoliative Glaucoma: Our Experience

Felicia M. Ferreri

Abstract

We investigated the correlation between hyperhomocysteinemia and pseudo-exfoliative glaucoma. After providing extensive background information, we outlined our study methodology. We assembled a control group of 20 individuals, considering their medical history (including hypertension, diabetes, cardiovascular and cerebrovascular diseases, nephropathies, and inappropriate drug therapy). Our study focused exclusively on patients with secondary open-angle glaucoma associated with pseudo-exfoliation, which is the most common cause of open-angle glaucoma. Our finding indicates that hyperhomocysteinemia is significantly elevated in subjects with pseudo-exfoliative glaucoma compared to individuals without ocular pathology but with a similar vascular risk. Homocysteine, by promoting the overproduction of free radicals, damages the intima of blood vessel walls and triggers elastase release in arterial smooth muscle cells. Antioxidants play a crucial role in mitigating the harmful effects of hyperhomocysteinemia, and folic acid supplementation, either alone or in combination with vitamins B12 and B6, improves endothelial function.

Keywords: glaucoma, pseudo-exfoliative syndrome, homocysteine, hyperhomocysteinemia, blindness

1. Introduction

Pseudo-exfoliative glaucoma is the most common cause of secondary glaucoma. Pseudo-exfoliative syndrome, from which it takes its name, is the main risk factor for the onset of glaucomatous disease, the second most common cause of visual impairment and blindness in Italy [1].

We refer the interested reader to Topouzis et al. [2] for a discussion of risk factors for glaucoma in a specific population; in addition, interested readers will find a detailed discussion of pseudo-exfoliative glaucoma in two very recent papers Washington et al. [3] and Yuksle et al. [4].

Homocysteine is an amino acid that plays a major role among the predisposing factors for thrombophilia, and the development of systemic and vascular pathologies and its role in pseudo-exfoliative glaucoma has been illustrated in [3].

Our study showed that there was a correlation between patients with pseudo-exfoliative glaucoma and the actual presence of hyperhomocysteinemia.

2. Background

2.1 Intraocular pressure (IOP)

The balance between aqueous humour production and absorption in the anterior chamber is crucial for pressure which usually ranges between 10 and 20 mmHg. In the glaucoma patient, this mechanism fails, hindering the outflow of aqueous humour, which, as it accumulates, causes an increase in intraocular pressure (IOP). This increase in pressure will result in mechanical damage at the head of the optic papilla. The damage caused by the increased pressure will also have a microvascular effect on the optic nerve with relative hypoperfusion and apoptosis of its nerve fibres. The loss of all nerve fibres will irreversibly result in blindness.

2.2 Glaucoma

Glaucoma is a very large group of ocular neuropathies characterised by progressive and irreversible damage to the optic nerve Gelormini et al. [5] and Caporossi et al. [6] for a general introduction and Flammer [7] for a more focused survey on glaucoma. The acute onset rarely occurs, often the course is subtle and progressive, resulting in functional damage to the visual field. In the past, the Von Graefe triad, which required a high IOP above 21 mmHg, functional visual field defects, and papilla neuropathy were used to detect glaucoma Flammer [7].

A number of studies have shown that many glaucomas do not have a high IOP and just as many have visual field damage, but only detected late.

In the face of these findings, of the previous triad, the only remaining stronghold is optic neuropathy Flammer [7].

Thus, one can speak of glaucoma considering optic nerve damage. An elevated IOP does not indicate pathology, nor can the presence of glaucoma lead to an elevated IOP.

2.3 Classification of glaucoma

We can distinguish glaucoma into two groups:

1. Clinically evident glaucoma, a much rarer manifestation with more or less obvious symptomatology and simply diagnosable among manifest papillary damage and high IOP. In this group, we include acute glaucoma, congenital glaucoma, and secondary, albeit more subtle and often chronic glaucoma.

2. Clinically silent glaucoma, much more common than the former.

We can also classify glaucoma according to the correlation of another ocular pathology thus distinguishing:

1. Primary glaucoma when there is no concomitant second pathology.

2. Secondary glaucoma when it is accompanied by a different ocular pathology.

We can make use of a third, more commonly accepted classification, which is based on the pathophysiology of the aqueous humour and its outflow, recognising two different pathologies based on the chamber angle:

1. Open-angle glaucoma.

2. Angle-closure glaucoma.

There is no primary classification or one that is not taken into account, but they can intersect with each other. We can therefore have, for example, clinically evident, secondary, closed-angle glaucoma.

Open-angle glaucoma is the most common type of glaucoma in this category and is mainly influenced by genetic predisposition; risk factors such as diabetes or myopia can have a significant impact on the development of open-angle glaucoma Flammer [7]. Interactions of some genes with glaucoma have been discovered, among them the GLC 1A gene and the Duffy group, a blood marker, both on chromosome 1. Non-modifiable risk factors include older age and ethnicity. By the time an alteration of the visual field occurs, optic nerve atrophy is already advanced.

2.4 Pseudo-exfoliative glaucoma

Secondary open-angle glaucoma includes glaucoma due to pseudo-exfoliative syndrome (PEX syndrome). We refer the reader to Flammer, Kozart and Yanof, and Ritch [7–9] for an introduction on exfoliative syndrome. The association between exfoliation syndrome and glaucoma has been reviewed in Ritch [10] and Mitchel et al. [11]; we also refer the interested reader to Jeng et al. [12] for the results of a study carried out in a very specific population. More recently, some studies explored the genetic association between pseudo-exfoliative syndrome and glaucoma Schlötzer-Schrehardt [13], Thorleifsoon et al. [14] and Ye et al. [15]. In addition, Ritch [16] and Ritch [17] describe ties between exfoliation syndrome and open glaucoma; Jeng [12], Forsius [18], and Ringvold et al. [19] discuss empirical results obtained on some specific populations. However, we point out that the exfoliation syndrome has been associated with other type of diseases such as vascular diseases Praveen et al. [20], Mitchell et al. [21], Shrum et al. [22], Schumacher et al. [23], Alzheimer Linner et al. [24] and Cumurcu et al. [25], and to some extent, it can be regarded as a a systemic disorder [26–28].

PEX syndrome is a systemic extracellular matrix (ECM) disorder characterised by the production and gradual accumulation of fibrillar material, not only at the skin level, but also at the connective tissue level of various organs, leading to glaucoma and increased cerebrovascular and cardiovascular morbidity such as Alzheimer's disease [24, 25], hearing loss [29–31], and increased plasma levels of homocysteine that increase the risk of thrombotic events.

PEX was first defined by Lindberg in 1917 and associated with glaucoma only a few years later in 1924.

It represents the most common form of secondary open-angle glaucoma with approximately 70 million people suffering from PEX and of these about 10% develop chronic open-angle glaucoma.

This term, pseudo-exfoliative, was chosen because clinically it has been observed that the lens appears to exfoliate but this is not what happens during this disease. Some important studies focus on the clinical detection of exfoliation syndrome [32–35]; the variation of IOP in patients with exfoliation glaucoma and patients with primary open-angle glaucoma is extensively analysed in Konstas et al. [36], while Konstas et al. [37] discuss the factors influencing the progression of exfoliation glaucoma. Medical treatments for exfoliation glaucoma are presented in

[38–40]; biomarkers for exfoliation glaucoma are reviewed in [36, 41]. The role of connective tissue growth in exfoliation glaucoma is investigated by Browne et al. [42]; also humour aqueous plays an important role, as emerges from [43–46]. At the surgical level, some important studies are proposed in Shingleton et al. [47] and Yazgan et al. [48].

Since not all patients with PEX syndrome go on to develop glaucoma, a distinction must be made between PEX syndrome, PEX syndrome-associated glaucoma, and pseudo-exfoliative.

PEX syndrome presents grey/white deposits formed by an abnormal protein material found on any ocular surface wetted by aqueous humour. It is most easily seen on the anterior surface of the crystalline lens where it presents with a characteristic picture with a central disc and a peripheral band due to iris movements rubbing against the lens during mydriasis and miosis movements and enucleating parts of the deposited material.

This rubbing also generates a loss of pigment from the deep layers of the iris, which is highlighted on transillumination. The light that penetrates through the pupil is reflected by the fundus and usually exits through the iris, which by its structure acts as an optical diaphragm. This mechanism fails in areas where there has been a greater loss of pigment at the level of the iris.

The affected areas appear red due to the backlighting of the light reflection from the ocular fundus because they are highly vascularised to the point of calling the colour 'fundus red'.

This phenomenon also explains the reason for red eyes in photos taken with flash.

At the level of the anterior capsule of the lens, these deposits of furfuraceous material take on an identifiable arrangement in three zones: a central disc, a peripheral granular zone, and an intermediate ring without deposits.

It is also important that pseudo-exfoliative material is found at the level of the corneal Descemet's membrane, and instead at the level of the iris and trabecular meshwork where it causes an obstruction of aqueous humour outflow with a related increase in IOP, leading to the onset of pseudo-exfoliative glaucoma.

These deposits can cause other ocular pathologies, e.g. increased fragility of the zonular fibrils that hold the lens in place, leading to possible complications in cataract surgery.

PEX syndrome is an age-related condition, generally from the sixth decade of life and is more common than one would think and fibrillar material can also occur in other organs.

The fibrillar material appears as a bush and can be observed under light microscopy. The electron microscopic presentation of pseudo-exfoliative disease in ocular and extra-ocular tissues was demonstrated by Schlötzer-Schrehardt.

The aetiology of this disease is unknown but has been classically linked to an altered metabolism of elastin fibrils. This hypothesis was confirmed by a study that demonstrated polymorphisms in the gene for lysyl oxidase-like protein 1 (LOXL1) [15]. This enzyme is involved in elastin metabolism and confers an increased risk of developing PEX syndrome. Several variants were found related to the POMP, CACNA1A, and SEMA6A genes, all three of which are linked to extracellular matrix metabolism, the ubiquitin-proteasome system, calcium signalling, and lipid biosynthesis in the pathogenesis of pseudo-exfoliation, increasing the risk of disease.

It is common in Scandinavian and Mediterranean populations, African Bantus, and Australian Aborigines, and moreover, it is more common in females.

A distinction should be made between primary open-angle glaucoma and pseudo-exfoliative glaucoma: usually, in the pseudo-exfoliative type, the IOP increases very quickly with considerable fluctuations that may result in greater damage to the optic disc than a high IOP yes, but still stable. Glaucomatous damage is also represented by the vascular changes associated with PEX glaucoma.

Another alteration due to PEX syndrome is the reduction of tear secretion and tear film stability.

One study showed that the tear osmolarity in both eyes of patients with clinically unilateral PES is higher than in normal subjects.

Small deposits of subjects with pseudo-exfoliative disease deposit on the corneal endothelium pseudo-exfoliation cells with reduced density.

Damaged corneal endothelium can cause endothelial decompensation. Pleomorphism in pseudo-exfoliation keratopathy with glaucoma is more frequent than with cataracts.

Recent studies using optical coherence tomography angiography have demonstrated a decrease in peripapillary and macular vascular density in patients with pseudo-exfoliation, suggesting that the vascular component, including optic nerve hypoperfusion, may be involved in the aetiopathogenesis of pseudo-exfoliative glaucoma.

3. Homocysteine

Homocysteine is a non-protein amino acid produced by the metabolism of methionine, an essential sulphur amino acid that is introduced into the body via the protein in the diet. Homocysteine is formed from S-adenosylmethionine (SAM) and S-adenosylhomocysteine (SAH). The transformation from SAM to SAH occurs via trans-methylation processes.

The SAM/SAH ratio underlies the regulation of methionine-homocysteine metabolism. In a well-functioning organism, homocysteine is converted back into methionine, or into simple amino acids, which are easily eliminated from the body via the urine. Approximately 80% of homocysteine in the blood is bound to proteins, mainly albumin, via a disulphide bond. The unbound portion of homocysteine is easily oxidised to form disulphides: due to the presence of its free SH group, it can combine with another homocysteine to form the homocysteine dimer or with cysteine to form the mixed cysteine-homocysteine disulphide. The set of these non-protein complexed forms (20% of total homocysteine) is called *free homocysteine.* This includes, therefore, both the reduced (SH) and the oxidised form: (S–S).

The strongly endothelium-toxic reduced form (SH) is approximately 2% and increases when plasma total homocysteine exceeds 100 μmol/L.

3.1 Homocysteine remethylation

Homocysteine can be remethylated into methionine by two processes:

In the former, with the presence of folic acid, in the folate cycle, the key reaction takes place thanks to the enzyme methylene tetrahydrofolate reductase (MTHFR), the coenzyme of which is vitamin B2, which reduces 5,10-methylene-tetrahydrofolate to 5-methyltetrahydrofolate; the latter will then provide, again with vitamin B12, the methyl group necessary for the reconversion of homocysteine to methionine.

In the second process, the remethylation reaction is carried out by the enzyme betaine synthase, which produces methionine by catalysing the transfer of a methyl group to homocysteine.

3.2 Hyperhomocysteinemia

The test to analyse homocysteine values in the blood is called *homocysteinemia* and consists of a venous blood sample.

Before the examination, it is preferable to fast for 10–12 hours and avoid smoking for at least 15 minutes prior to taking the sample.

Plasma homocysteine values are considered normal when they are around 5–12 μmol/L.

Hyperhomocysteinemia occurs when values exceed the maximum cut-off.

Let us distinguish hyperhomocysteinemia:

- Mild when the plasma value remains in the range of 12–15 μmol/L;
- Moderate when the range is 15–30 μmol/L;
- intermediate when between 30 and 100 μmol/L;
- Severe hyperhomocysteinemia when it exceeds the maximum cut-off.

The plasma concentration of homocysteine is the result of a close relationship between dietary habits and predisposing genetic factors.

Most people have high levels of homocysteine in their blood due to a diet that is not sufficiently rich in folic acid and the other B vitamins.

Other causes of hyperhomocysteinemia are genetic alterations that cause deficits in the enzymes involved in the metabolic cascade whose intermediate product is homocysteine.

In addition to deficiency of the enzyme cystathionine-beta-synthetase, due to a very rare genetic mutation, homocystinuria, elevated homocysteine levels may also be due to mutation of the gene responsible to produce the enzyme methylenetetrahydrofolate reductase (MTHFR). A genetic polymorphism has been identified as responsible for the increased homocysteine levels, characterised by the 1298A/C and C677T mutations; with the latter being more important in terms of thrombotic risk, resulting in a 50% reduction in MTHFR enzymatic activity.

Genetic causes outnumber dietary/behavioural ones 1:10.

3.3 Risks of hyperhomocysteinemia

The patient with hyperhomocysteinemia is an individual with a high predisposition to thrombophilia and thus to cardiovascular disease.

Homocysteine has been shown to influence vascular function through indirect action on muscle tone by inducing increased vascular constriction mediated by the binding of reduced homocysteine with nitric oxide and related nitrous oxide formation [49, 50]. Several researchers explored the role of homocysteine and acute coronary syndromes [50], abdominal aortic aneurysm [51], atherosclerosis [52], retinal vein occlusion [53], diabetic retinopathy [54–56], and ischemic strokes [57, 58]. Other important studies focus, instead, on the impact of some drugs and foods on homocysteine levels [59–64].

Chronically elevated homocysteine levels result in nitric oxide depletion, with nitrous oxide production remaining in the circulation for only 14 minutes. The resulting consequence is the patient in continuous vasospasm. A direct influence of high serum homocysteine levels causes atherosclerotic plaque formation and smooth muscle cell proliferation due to endothelial damage and reduced elasticity. This is due to excess homocysteine forming the homocysteine-thiolactone complex, which reacts with LDL to generate an insoluble LDL-thiolactone complex. This complex is phagocytosed by macrophages which, unable to metabolise it, turn into foamy cells acting as an atheromatous core.

Excess homocysteine can also act as a free oxygen radical causing first endothelial dysfunction and then necrosis of the endothelial cells themselves with their subsequent detachment from the vessel wall. There may also be a proliferation of smooth muscle cells resulting in fibrocalcification of the vessel wall and oxidation of membrane lipids with loss of function of these structures.

Again, related to increased thrombotic risk, excessive homocysteine is also a strong platelet aggregator.

3.4 Therapy

Therapy is based first and foremost on a correction of the patient's lifestyle by making changes to the diet.

In fact, the body can decrease serum homocysteine levels by means of folates and B vitamins, which are crucial in its metabolism.

It is essential for the patient to consume more food with folic acid and vitamin B12: the recommended intake levels for folic acid are 200–1000 mcg/day.

4. Correlation between pseudo-exfoliative glaucoma and hyperhomocysteinemia

4.1 Markers

Corneal biomechanical properties of subjects with pseudo-exfoliative syndrome, including corneal hysteresis, corneal resistance factor, and central corneal thickness, were reported to be decreased compared to healthy control subjects.

These changes were more pronounced in patients with pseudo-exfoliative glaucoma than in patients with pseudo-exfoliative syndrome.

Two interesting systemic laboratory markers for predicting the risk of progression from pseudo-exfoliative syndrome to pseudo-exfoliative glaucoma are the neutrophil to lymphocyte ratio and the platelet to lymphocyte ratio. Indeed, signs of subclinical systemic inflammation, including neutropenia and lymphocytopenia, became more pronounced in patients with PEX syndrome and even more so in those with pseudo-exfoliative glaucoma than in normal healthy subjects.

This finding suggests that inflammation plays a key role from the onset of PEX syndrome and can be used as a marker at follow-up to predict progression to pseudo-exfoliative glaucoma.

Chronic subclinical inflammation was also demonstrated locally in the anterior chamber.

Levels of activation-derived complement components were significantly elevated in the aqueous humour of patients with pseudo-exfoliative glaucoma compared to non-syndromic PEX glaucoma controls.

Complement inhibitors, including clusterin and vitronectin, were significantly elevated.

The association of clusterin with exfoliation fibrils suggests an unsuccessful attempt by this chaperonin to prevent the accumulation of fibrillar material. Apolipoprotein D (ApoD) is a secreted glycoprotein with multiple functions, including the prevention of lipid peroxidation. ApoD expression is upregulated during ageing and in several pathological conditions, including atherosclerosis, neurological diseases, and several types of cancer. Several studies have suggested that ApoD plays a protective role against oxidative stimuli.

This role was found to be diminished in PEX syndrome, where lower levels of this biomarker were detected in the aqueous humour of patients with pseudo-exfoliative syndrome.

Endothelin-1 is a peptide produced by the vascular endothelium with potent vasoconstrictor properties.

It appears to be involved in the regulation of ocular blood flow. High levels of endothelin-1 have been reported in the aqueous humour of patients with PEX syndrome. High levels of homocysteine have been found in both aqueous humour and tears.

High levels of this non-protein amino acid contribute to the pathogenesis of pseudo-exfoliative disease.

4.2 Purpose of the study

Both PEX and hyperhomocysteinemia are risk factors for systemic vascular and ocular disease.

We studied the correlation of hyperhomocysteinemia in patients with pseudo-exfoliative glaucoma compared to a healthy control group.

4.3 Methods

For each patient, we considered her/his medical history (hypertension, diabetes, cardiovascular and cerebrovascular diseases, nephropathies, and inappropriate drug therapy).

The control group was formed by 20 individuals without the disease who underwent a complete eye examination, including fundus and optic disc examination, visual field analysis, and other routine examinations to detect any elements that might affect the study.

We only included patients with secondary open-angle glaucoma with pseudo-exfoliation, the most common cause of open-angle glaucoma.

The study group consisted of glaucomatous patients diagnosed with pseudo-exfoliative glaucoma if clinical examination revealed fibrillar deposits on the anterior lens capsule and at the level of the ciliary angle with associated peripupillary atrophy and transillumination of the pupillary margin, elevated intraocular pressure, incipient optic nerve atrophy, and glaucomatous visual field defects.

Patients were excluded if they did NOT have a diagnosis of pseudo-exfoliative glaucoma based on clinical examination and visual field analysis, if they had diseases that could be associated with hyperhomocysteinemia, such as gastrointestinal malabsorption or diabetes mellitus, if they were abusing drugs and/or alcohol, and if they were receiving any pharmacological treatment, including methotrexate therapy and other possible vitamin supplements, in the 6 months prior to the study. Patients

DOI: http://dx.doi.org/10.5772/intechopen.1005897

	No. of subjects	Age (in years)	Homocysteinemia	Hyperhomo-cysteinemia (%)
Study group	20	65–75 70 ± 5.2	16.6 ± 3.1	50
Control group	20	64–76 70 ± 6.1	12.41 ± 1.8	10

Table 1.
Features of the patients involved in our study.

with glaucoma and on medical therapy were not excluded, as no glaucoma therapy is known to alter serum homocysteine levels.

For each participant in both the control and study groups, a 2 ml venous blood sample was collected and centrifuged for 6 minutes using the fluorescence-based high-performance liquid chromatography method.

A plasma value above 16 μmol/L is considered high.

The control group consisted of the same number of healthy adults, 20, again divided into 10 men and 10 women, with a mean age close to that of the study group, ranging from 64 to 76 years (mean 70 ± 6.1).

It was immediately apparent that the mean plasma homocysteine level was significantly higher in the pseudo-exfoliative glaucoma group.

Fifty percent of the patients in the pseudo-exfoliative glaucoma group had plasma levels to report hyperhomocysteinemia; in the control group the number was drastically lower: only 2/20 (10%).

Various cardiovascular diseases were found in the study group. For example, four patients (20%) had systemic hypertension, and two (10%) had ischaemic heart disease.

The two patients with pseudo-exfoliative glaucoma and ischaemic heart disease had elevated plasma homocysteine levels, as did 3 of the 4 patients with systemic hypertension.

In the control group, cardiovascular disease was found in four patients with systemic hypertension and one patient with ischaemic heart disease.

The patient with ischaemic heart disease and one of the four patients with hypertension had hyperhomocysteinemia (**Table 1**).

5. Discussion

We found that the prevalence of hyperhomocysteinemia is significantly increased in subjects with pseudo-exfoliative glaucoma compared to subjects without ocular pathology but with a similar vascular risk.

Homocysteine can induce overproduction of free radicals which, by causing damage to the intima of the vessel wall, trigger elastase at the level of arterial smooth muscle cells.

This activation leads to elastolysis of elastin and fibrillar collagen in the arteries through activation of extracellular matrix metalloprotease and may explain the possible effect of homocysteine in systemic cardiovascular disease.

Antioxidants such as vitamin E and ascorbic acid are able to generate a significant reduction in the harmful effects of hyperhomocysteinemia.

Treatment supported by folic acid, which is able to eliminate free oxygen radicals, led to a general improvement in endothelial function, alone or in combination with the intake of B vitamins (B12 and B6).

6. Conclusions

The pseudo-exfoliative material contains both elastin and fibrin, and homocysteine is able to activate an elastase inducing elastolysis of elastin and fibrillar collagen.

This correlation could highlight new developments to better understand how fibrillar material is produced in pseudo-exfoliative disease and consequently take a step forward in the screening of pseudo-exfoliative glaucoma and the plausible role homocysteine plays in the pathogenesis of vascular disease in both the systemic and ocular spheres.

Indeed, there is a possibility, yet to be confirmed, that the pseudo-exfoliative material generated is actually a by-product of homocysteine-induced elastolysis.

A further line of development could be to evaluate the role of folic acid in pseudo-exfoliative glaucoma patients with hyperhomocysteinemia on IOP.

Author details

Felicia M. Ferreri
BIOMORF Department, University of Messina, Messina, Italy

*Address all correspondence to: fferreri@unime.it

References

[1] Rossetti L, Digiuni M, Giovanni M, Centofanti M, Fea AM, Iester M, et al. Blindness and glaucoma: A multicenter data review from 7 academic eye clinics. PLoS One. 2015;**10**(8):e0136632

[2] Topouzis F, Wilson MR, Harris A, Founti P, Yu F, Anastasopoulos E, et al. Risk factors for primary open-angle glaucoma and pseudoexfoliative glaucoma in the Thessaloniki eye study. American Journal of Ophthalmology. 2011;**152**:219-212

[3] Washington J, Ritch R, Liu Y. Homocysteine and glaucoma. International Journal of Molecular Sciences. 2023;**24**(13):10790

[4] Yüksel N, Tuğan BY. Pseudoexfoliation glaucoma: Clinical presentation and therapeutic options. Turkish Journal of Ophthalmology. 2023;**53**(4):247

[5] Gelormini F, Querques Q, Guandolfi S. Handbook of general ophthalmology, Saluzzo. Minerva Medica. 2022:103-116

[6] Caporossi A, Salgarcllo T, Manganelli C. Ophthalmology. 2017;**201**:223-233

[7] Flammer J. Glaucoma, a guide for patients, an introduction for physicians, a reference for quick reference. Verduci. 2001:63-65

[8] Kozart DM, Yanoff M. Intraocular pressure status in 100 consecutive patients with exfoliation syndrome. Oftalmologia. 1982;**89**:214-218

[9] Ritch R. Exfoliation syndrome. Current Opinion in Ophthalmology. 2001;**12**:124-130

[10] Ritch R. Perspective on exfoliation syndrome. Journal of Glaucoma. 2001;**10**(5 Suppl. 1):S33-S35

[11] Mitchell P, Wang JJ, Hourihan F. The relationship between glaucoma and pseudoexfoliation: The blue mountains eye study. Archives of Ophthalmology. 1999;**117**:1319-1324

[12] Jeng SM, Karger RA, Hodge DO, Burke JP, Johnson DH, Good MS. The risk of glaucoma in pseudoexfoliation syndrome. Journal of Glaucoma. 2007;**16**:117-121

[13] Schlötzer-Schrehardt U. Molecular pathology of pseudoexfoliation/glaucoma syndrome-new insights from LOXL1 gene associations. Experimental Eye Research. 2009;**88**:776-785

[14] Thorleifsson G, Magnusson KP, Sulem P, Walters GB, Gudbjartsson DF, Stefansson H, et al. Common sequence variants in the LOXL1 gene confer susceptibility to exfoliative glaucoma. Science. 2007;**317**:1397-1400

[15] Ye H, Jiang Y, Jing Q, et al. LOXL1 hypermethylation in pseudoexfoliation syndrome in the Uyghur population. Investigative Ophthalmology & Visual Science. 2015;**56**:5838-5843

[16] Ritch R, Schlötzer-Schrehardt U, Konstas AG. Why is glaucoma associated with exfoliation syndrome? Progress in Retinal and Eye Research. 2003;**22**:253-275

[17] Ritch R. Exfoliation syndrome: The most common identifiable cause of open-angle glaucoma. Journal of Glaucoma. 1994;**3**:176-177

[18] Forsius H. Exfoliation syndrome in various ethnic populations. Acta

Ophthalmologica. Supplement. 1985;**1988**(184):71-85

[19] Ringvold A, Blika S, Elsås T, Guldahl J, Brevik T, Hesstvedt P, et al. The central Norwegian eye screening study. I. Epidemiology of pseudoexfoliation syndrome. Acta Ophthalmologica. 1988;**66**:652-658

[20] Praveen MR, Shah SK, Vasavada AR, Diwan RP, Shah SM, Zumkhawala BR, et al. Pseudoexfoliation as a risk factor for peripheral vascular disease: A case-control study. Eye (London, England). 2011;**25**:174-179

[21] Mitchell P, Wang JJ, Smith W. Association of pseudoexfoliation syndrome with increased vascular risk. American Journal of Ophthalmology. 1997;**124**:685-687

[22] Shrum KR, Hattenhauer MG, Hodge D. Cardiovascular and cerebrovascular mortality associated with ocular pseudoexfoliation. American Journal of Ophthalmology. 2000;**129**:83-86

[23] Schumacher S, Schlotzer-Schrehardt U, Martus P, et al. Pseudoexfoliation syndrome and abdominal aortic aneurysms. Lancet. 2001;**357**:359-360

[24] Linnér E, Popovic V, Gottfries CG, Jonsson M, Sjögren M, Wallin A. Exfoliation syndrome in cognitive impairment of the cerebrovascular or Alzheimer's type. Acta Ophthalmologica Scandinavica. 2001;**79**(3):283-285

[25] Cumurcu T, Dorak F, Cumurcu BE, Erbay LG, Ozsoy E. Is there any relationship between pseudoexfoliation syndrome and Alzheimer's type dementia? Seminars in Ophthalmology. 2013;**28**:224-229

[26] Schlötzer-Schrehardt UM, Koca MR, Naumann GO, Volkholz H, Pseudoexfoliation syndrome. Ocular manifestation of a systemic disorder? Archives of Ophthalmology. 1992;**110**:1752-1756

[27] Schlötzer-Schrehardt U, Naumann GO. Ocular and systemic pseudoexfoliation syndrome. American Journal of Ophthalmology. 2006;**141**:921-937

[28] Ritch R, Schlotzer-Schrehardt U. Exfoliation syndrome. Survey of Ophthalmology. 2001;**45**:265-315

[29] Aydoğan Ozkan B, Yüksel N, Keskin G, Altintaş O, Karabaş VL, Cağlar Y, et al. Plasma homocysteine levels and sensorineural hearing loss in patients with pseudoexfoliation syndrome. European Journal of Ophthalmology. 2006;**16**:542-547

[30] Cahill M, Early A, Stack S, Blayney AW, Eustace P. Pseudoexfoliation and sensorineural hearing loss. Eye (London, England). 2002;**16**:261-266

[31] Singham NV, Zahari M, Peyman M, Prepageran N, Subrayan V. Association between ocular pseudoexfoliation and sensorineural hearing loss. Journal of Ophthalmology. 2014;**2014**:825936

[32] Prince AM, Ritch R. Clinical signs of pseudoexfoliation syndrome. Ophthalmology. 1986;**93**:803-807

[33] Köse HC, Tekeli O. Optical coherence tomography angiography of the peripapillary region and macula in normal, primary open-angle glaucoma, pseudoexfoliation glaucoma and ocular hypertension eyes. International Journal of Ophthalmology. 2020;**13**:744-754

[34] Cornelius A, Pilger D, Riechardt A, Reitemeyer E, Rübsam A, Winterhalter S,

et al. Macular, papillary and peripapillary perfusion densities measured by optical coherence tomography angiography in primary open-angle glaucoma and pseudoexfoliation glaucoma. Graefe's Archive for Clinical and Experimental Ophthalmology. 2022;**260**:957-965

[35] Subasi S, Yuksel N, Basaran E, Pirhan D. Comparison of vessel density in macular and peripapillary regions between primary open-angle glaucoma and pseudoexfoliation glaucoma using OCTA. International Ophthalmology. 2021;**41**:173-184

[36] Konstas AG, Hollo G, Astakhov YS, Teus MA, Akopov EL, Jenkins JN, et al. Factors associated with progression or long-term stability in exfoliation glaucoma. Archives of Ophthalmology. 2004;**122**:29-33

[37] Konstas AG, Mantziris DA, Stewart WC. Diurnal intraocular pressure in untreated exfoliation and primary open-angle glaucoma. Archives of Ophthalmology. 1997;**115**:182-185

[38] Konstas AG, Kozobolis VP, Katsimpris IE, Boboridis K, Koukoula S, Jenkins JN, et al. Efficacy and safety of latanoprost compared with travoprost in patients with exfoliative glaucoma. Ophthalmology. 2007;**114**:653-657

[39] Parmaksiz S, Yuksel N, Karabas VL, Ozkan B, Demirci G, Caglar Y. A comparison of travoprost, latanoprost and the fixed combination of dorzolamide and timolol in patients with pseudoexfoliation glaucoma. European Journal of Ophthalmology. 2006;**16**:73-80

[40] Yüksel N, Gök M, Altıntaş O, Cağlar Y. Diurnal intraocular pressure efficacy of timolol-brimonidine fixed combination and timolol-dorzolamide fixed combination as first-choice therapy in patients with pseudoexfoliation glaucoma. Current Eye Research. 2011;**36**:804-808

[41] McNally S, O'Brien CJ. Metabolomics/proteomics strategies used to identify biomarkers for exfoliation glaucoma. Journal of Glaucoma. 2014;**23**(suppl. 1):S51-S54

[42] Browne JG, Ho SL, Kane R, et al. Connective tissue growth factor is increased in pseudoexfoliation glaucoma. Investigative Ophthalmology & Visual Science. 2011;**52**:3660-3666

[43] Ho SL, Dogar GF, Wang J, et al. Elevated aqueous humour tissue inhibitor of matrix metalloproteinase-1 and connective tissue growth factor in pseudoexfoliation syndrome. The British Journal of Ophthalmology. 2005;**89**:169-173

[44] Gayathri R, Coral K, Sharmila F, et al. Correlation of aqueous humour lysyl oxidase activity with TGF-ss levels and LOXL1 genotype in pseudoexfoliation. Current Eye Research. 2016;**41**:1331-1338

[45] Klinchnikova AA, Samokhina NI, Ilina IY, et al. Proteome of the human aqueous humour in cataract, glaucoma and pseudoexfoliation syndrome. Proteomics. 2016;**16**:131938-131946

[46] Koliakos GG, Konstas AG, Schlötzer-Schrehardt U, et al. Endothelin-1 concentration is increased in the aqueous humour of patients with exfoliation syndrome. The British Journal of Ophthalmology. 2004;**88**:523-527

[47] Shingleton BJ, Crandall AS, Ahmed II. Pseudoexfoliation and cataract surgery: Preoperative, intraoperative and postoperative issues related to intraocular pressure, cataract and intraocular lenses. Journal of Cataract and Refractive Surgery. 2009;**35**:1101-1120

[48] Yazgan S, Celik U, Alagöz N, et al. Corneal biomechanical comparison of pseudoexfoliation syndrome, pseudoexfoliative glaucoma and healthy subjects. Current Eye Research. 2015;**40**:470-475

[49] Hankey GJ, Eikelboom JW. Homocysteine and vascular disease. Lancet. 1999;**357**:407-413

[50] Stubbs PJ, Al-Obaidi MK, Conroy RM, et al. Effect of plasma homocysteine concentration on early and late events in patients with acute coronary syndromes. Circulation. 2000;**102**:605-610

[51] Tribouilloy CM, Peltier M, Iannetta Peltier MC, et al. Plasma homocysteine and severity of thoracic aortic atherosclerosis. Chest. 2000;**118**:1685-1689

[52] Brunelli T, Prisco D, Fedi S, et al. High prevalence of mild hyperhomocysteinemia in patients with abdominal aortic aneurysm. Journal of Vascular Surgery. 2000;**32**:531-536

[53] Grapevine AK. Hyperhomocysteinemia: A risk factor for central retinal vein occlusion. American Journal of Ophthalmology. 2000;**129**:640-644

[54] Pianka P, Almog Y, Man O, et al. Hyperhomocysteinemia in patients with non-arteritic anterior ischemic optic neuropathy, central retinal artery occlusion and central retinal vein occlusion. Ophthalmology. 2000;**107**:1588-1592

[55] Neugebauer S, Baba T, Kurokawa K, et al. Defective homocysteine metabolism as a risk factor for diabetic retinopathy. Lancet. 1997;**349**:473-474

[56] Cursiefen C, Tammer T, Kuchle M, et al. Pseudoexfoliation syndrome in eyes with ischemic central retinal vein occlusion. A histopathological and electron microscopic study. Acta Ophthalmologica Scandinavica. 2001;**79**:476-478

[57] Hoogeveen EK, Kostense PJ, Eysink PE, et al. Hyperhomocysteinemia is associated with the presence of retinopathy in type 2 diabetes mellitus: The Hoorn study. Archives of Internal Medicine. 2000;**160**:2984-2990

[58] Penix LP. Ischemic strokes secondary to vitamin B12 deficiency-induced hyperhomocystinemia. Neurology. 1998;**51**:622-624

[59] Hankey GJ, Eikelboom JW. Homocysteine and stroke. Current Opinion in Neurology. 2001;**14**:95-102

[60] Guttormsen AB, Ueland PM, Lonning PE, et al. Kinetics of plasma total homocysteine in patients on high-dose methotrexate therapy. Clinical Chemistry. 1998;**44**:1987-1989

[61] Dierkes J, Westphal S, Luley C. Serum homocysteine increases after fenofibrate or bezafibrate therapy. Lancet. 1999;**354**:219-220

[62] Westphal S, Dierkes J, Luley C. Effects of fenofibrate and gemfibrozil on plasma homocysteine. Lancet. 2001;**358**:39-40

[63] van der Gaag MS, Ubbink JB, Sillanaukee P, et al. Effect of red wine, liquor and beer consumption on serum homocysteine. Lancet. 2000;**355**:1522

[64] Fiskerstrand T, Refsum H, Kvalheim G, et al. Homocysteine and other thiols in plasma and urine: Automated determination and sample stability. Clinical Chemistry. 1993;**39**:263-271